SAP UI Frameworks for Enterprise Developers

A Practical Guide

Marius Obert
Volker Buzek
Foreword by Peter Muessig

Apress®

SAP UI Frameworks for Enterprise Developers: A Practical Guide

Marius Obert
Munich, Germany

Volker Buzek
Bielefeld, Germany

ISBN-13 (pbk): 978-1-4842-9534-2
https://doi.org/10.1007/978-1-4842-9535-9

ISBN-13 (electronic): 978-1-4842-9535-9

Managing Director, Apress Media LLC: Welmoed Spahr
Acquisitions Editor: Divda Modi
Development Editor: Laura Berendson
Editorial Assistant: Divda Modi

Cover designed by eStudioCalamar

Distributed to the book trade worldwide by Springer Science+Business Media New York, 1 New York Plaza, Suite 4600, New York, NY 10004-1562, USA. Phone 1-800-SPRINGER, fax (201) 348-4505, e-mail orders-ny@springer-sbm.com, or visit www.springeronline.com. Apress Media, LLC is a California LLC and the sole member (owner) is Springer Science + Business Media Finance Inc (SSBM Finance Inc). SSBM Finance Inc is a **Delaware** corporation.

For information on translations, please e-mail booktranslations@springernature.com; for reprint, paperback, or audio rights, please e-mail bookpermissions@springernature.com.

Apress titles may be purchased in bulk for academic, corporate, or promotional use. eBook versions and licenses are also available for most titles. For more information, reference our Print and eBook Bulk Sales web page at http://www.apress.com/bulk-sales.

Any source code or other supplementary material referenced by the author in this book is available to readers on GitHub (github.com/apress). For more detailed information, please visit http://www.apress.com/source-code.

Printed on acid-free paper

This book is dedicated to all open source developers out there. Your contributions have a significant impact on the world, and they inspire us. Keep going strong!

Volker would like to thank his wife Stella for her support throughout the writing process – ad astra, Puzzlestück! – and his kids for making time when needed (although they now know more audio books than they should).

Table of Contents

About the Authors

 **Marius Obert** started his software development career as a UI developer in sunny California at SAP Labs. During this time, he learned to love web technologies in general, such as JavaScript, and the entire Node.js ecosystem in particular.

After many ugly fights against CSS, he relocated to Munich, working as a Developer Evangelist. In this role, he tries to inspire his peer developers around the globe to build cloud-native apps with excellent customer engagement.

Besides his day job, Marius tries to keep current with Web 3.0 technologies.

 **Volker Buzek** works as a development architect in the SAP web/mobile cosmos, often at the boundary of SAP/non-SAP. As an SAP Mentor, he fosters the adoption of open source within the SAP ecosystem and is actively involved in the SAP community. In his spare time, the trained scuba instructor enjoys spending time at and in the ocean with his family.

About the Technical Reviewer

 **Peter Muessig**, chief architect at SAP, is responsible for several UI technologies such as SAPUI5, UI5 Web Components, Fundamental Library, and Luigi. His journey started in 2005 in SAP NetWeaver development. In 2008, he played a key role in evaluating different UI technology options. This evaluation led to the birth of Project Phoenix in 2009, an ambitious initiative to develop a modern, responsive, enterprise-grade web UI framework for business applications, ultimately resulting in SAPUI5.

Since then, Peter has made significant contributions across various areas within the SAPUI5 framework and its accompanying tools. His unwavering passion for change and innovation has kick-started projects such as SAP Web IDE, UI5 Evolution, UI5 Web Components, UI5 Tooling, and its open source ecosystem. Peter likes to explore other frameworks and technologies, finding value in both their practical application and their ability to inspire him with fresh ideas. In addition to his passion for UI, he deeply cares about the development experience and enjoys full-stack end-to-end development.

In 2022, Peter took on additional responsibilities for Fundamental Library and Luigi, focusing on aligning the different technology options to ensure compatibility and avoid redundancy. His goal is to provide an interoperable and complementary UI technology offering to cover all kinds of scenarios.

Acknowledgments

We would like to express our appreciation to all those who have patiently guided us in learning programming, whether in the classroom, at work, online, or elsewhere. This book owes much to your contributions, and we hope that it can serve as our humble offering to the community by sharing our knowledge.

A virtual wave goes to the global UI5 community! We hope that this book fosters it as the friendly, curiosity-driven place it was from the beginning.

We also extend our sincere gratitude to the Apress team for their valuable assistance and cooperation in the making of this book.

Foreword

Are you currently facing the challenge of making technology decisions for your project? If not, you've likely encountered this situation before or will do so in the future. As the chief architect at SAP for the SAP UI technologies OpenUI5/SAPUI5, UI5 Web Components, and Fundamental Library Styles, I often assist teams and customers in their UI technology decisions. It often comes down to personal preference, skills, requirements, and environment. That's why books like this one are invaluable for making independent technological decisions. They provide a quick overview and help you dive into technologies because there's nothing more important than forming your own impression.

I met Volker Buzek for the first time in 2018 at UI5con, the conference dedicated to UI5. Since then, we've stayed in close contact and occasionally meet up for hackathons. Together we started the UI5 ecosystem showcase to demonstrate the openness and extensibility of the UI5 Tooling and to provide several examples for the UI5 community.

At around the same time, I also connected with Marius Obert, who worked as a UI developer and Developer Advocate in cloud programming for SAP. Due to his proximity to UI, we had several connection points in the past and especially with the start of the easy-ui5 project, a project generator for UI5.

Last year, Volker approached me with the idea of authoring a book serving as a practical guide about the various SAP UI technologies and how to use them to build enterprise-grade web applications that follow the SAP Fiori design guidelines. I really liked the idea because there are hardly any books that compare these UI technologies and highlight their differences. In my opinion, it's essential to understand these differences, their value propositions, and the target environments they're designed for. Each SAP UI technology is tailored for specific use cases: Fundamental Library Styles, for example, is a lightweight presentation layer to style web pages and applications, whereas UI5 Web Components is a library providing independent and reusable UI elements. Both technologies are based on web standards and can therefore be used with any framework of your choice.

Frameworks themselves are coming with distinct feature sets and different levels of abstraction to shield developers from complexities and let them focus on their work: in OpenUI5/SAPUI5, you mainly work with predefined layouts and controls that you stitch together and share through views or components. The framework offers built-in data binding, allowing easy connection to backend services via dedicated models – a convenient solution to quickly build business applications. On top, you find SAP Fiori elements, which uses a template- and metadata-driven approach. With this framework, developers can easily create SAP Fiori applications based on OData services and annotations that don't need JavaScript UI coding at all.

As you can see, each of these technologies has advantages and disadvantages. Higher abstraction levels offer more convenience but less freedom. Unfortunately, there is no "one technology to rule them all," but it's essential to know which use cases each UI technology excels in.

Now it's your turn to dive into this book. Despite my own background and experience with these technologies, this book has provided me with valuable insights from an external perspective. Reading the chapters and following the concise and clear examples in this book has even inspired me with ideas on how to enhance these technologies to better meet your needs as consumers of the SAP UI technologies.

Finally, it's on you to decide: learn carefully, explore deeply, and choose wisely!

Peter Muessig

Introduction

Over the past decade, we've worked on countless projects that involved SAP UI technologies. And there was one constant throughout all the projects: the developers had the feeling that they needed to fight against the stigma that an SAP user interface cannot look nice and have a great user experience. While there are undoubtedly user interfaces with an unintuitive user experience, there are many examples of great user interfaces built with modern web technology suited for the task at hand. With the right tools and technologies for your use case, you can build the user interfaces your customer will love.

Chapter 1 of this book will introduce you to the user experience concept of SAP Fiori, which can be implemented with several state-of-the-art web technologies. Each following chapter will provide an overview of one of these popular technologies and explain in which situations this technology is beneficial. The hands-on tutorials in each will help you kick-start your development journey, from setting up your coding editor to building your first application with SAPUI5, SAP Fiori elements, UI5 Web Components, or Fundamental Library Styles. We know that technology, especially UI technology, can change rapidly. Hence, Chapter 5 will introduce you to new technologies that we already see on the horizon today and put them in the larger context of industry-wide trends. For trends beyond the horizon, Chapter 6 will point you to resources on the Web and communities in real life that will help you stay on top of future innovations. After reading this book, you will be able to build intuitive applications with appealing user interfaces that are SAP Fiori conform. **Modern web technologies and enterprise user interfaces can go hand in hand.**

Whom This Book Is For

This book is suitable for both junior and experienced developers. For junior developers, it serves as an introduction to enterprise software development and provides a comprehensive understanding of the requirements for building intuitive user interfaces for SAP applications. For experienced developers, the book covers new and emerging technologies in the past decade, as well as those on the horizon. Ultimately, the book is a comprehensive guide to building intuitive and visually appealing user interfaces for SAP applications using modern web technologies.

CHAPTER 1

Enterprise User Interfaces

In this chapter, we'll explore the key differences between user interfaces from enterprise applications and those found in consumer-grade apps. We'll also emphasize the importance of considering end users' preferences and needs when designing enterprise applications and provide insights on the tooling needed to build visually appealing user interfaces. Plus, we'll introduce you to the SAP Fiori design system and show you how it can be leveraged to create intuitive and efficient user interfaces for enterprise applications. Get ready to take your frontend skills to new heights and learn how to create interfaces that stand out.

The SAP Journey in a Nutshell

SAP has built quite a reputation in its more than 50-year history and is now considered the de facto standard software in many industries. The company successfully established itself as a global leader due to several factors. First, SAP provides comprehensive functionality and offers a wide range of Enterprise Resource Planning (ERP) modules and applications that support various business processes and functions, such as financial management, supply chain management, and human resources management. This functionality enables SAP to meet the needs of various businesses and industries. Second, SAP software offers a high

© Marius Obert and Volker Buzek 2023
M. Obert and V. Buzek, *SAP UI Frameworks for Enterprise Developers*,
https://doi.org/10.1007/978-1-4842-9535-9_1

degree of customization and flexibility, which allows businesses to tailor the software to meet their specific needs and requirements. As a result, SAP has received numerous awards and accolades for its ERP software, which has helped to establish its reputation as a leading provider of ERP solutions.

SAP also provides comprehensive support and maintenance services to its customers to help ensure that their systems are running smoothly and efficiently. Furthermore, the company has a global presence and established trusted partnerships with a wide range of companies and organizations around the world. This global reach and partnerships have contributed to SAP's success as a leader in the ERP market. Additionally, the enterprise has a strong focus on innovation and has invested significantly in research and development to continuously improve its products and services for the future of work.

And finally, SAP is also invested in strong branding and marketing. The corporation has effectively marketed its products and services to potential customers around the world. This has helped SAP to build a large customer base and establish itself as a leader in the ERP market. And history has proven that this was the right path.

To a Hammer, Everything Looks like a Nail

Independent of the commercial success, many users associate SAP's brand with a nonintuitive user experience and confusing user interfaces. Some might even have a negative sentiment toward SAP user interfaces in general. We've heard this multiple times throughout our careers, but it's difficult to quantify such statements. SAP is a large and complex enterprise software platform with a wide range of functionality and features, and the user interface design of individual SAP modules and applications may vary significantly. Some SAP user interfaces may be considered more aesthetically pleasing than others, and user interface design is often a personal preference.

But generally, there is a point to this criticism, and especially SAP's older user interfaces (such as the SAP GUI) look outdated today. It's important to note that most SAP R/3 user interfaces were written through the 1990s. Back then, object-oriented programming was not nearly as well understood as it is today, and it still took more than a decade until books like *Clean Code* were conceptualized. Cloud computing was unknown to the entire world, and the Internet was just about to take off. So naturally, A/B testing was not yet related to software development, and user research meant talking to a handful of end users. It's no surprise that the primary goal of a user interface has been to facilitate the use of the software and enable users to efficiently and effectively complete their tasks. The appearance of the user interface was a secondary concern compared to its functionality. As such, the design of a user interface may prioritize functionality over aesthetics, and there was only one tool to implement user interfaces: SE80 – don't worry if this doesn't mean anything to you.

Luckily, SAP started to address this issue a few years ago and has made much progress. SAP has adopted the design thinking approach, a human-centered design process that helps organizations understand and solve complex problems. The software company now uses design thinking to develop and improve its products and services and also offers design thinking training to its customers and partners to educate their ecosystem as well. Furthermore, SAP built a dedicated UX (user experience) strategy team of designers, researchers, and UX engineers, led by their Chief Design Officer. This team focuses on improving the user experience of SAP products and services and regularly works with customers and partners to understand their needs and pain points. Equipped with the right knowledge and tools, they design and develop solutions to address these needs. An important output of this team is the **SAP Fiori design system** that aims to provide a consistent and modern look and feel across all SAP applications. SAP Fiori provides a personalized, responsive, and simple UI for users, focusing on task-based and role-based interactions.

This design concept has proven itself in multiple applications across lines of businesses and industries and has been implemented in all of SAP's user interface technologies. Yes, that's right. The company **no longer bets on a single technology** that is being forced on all developers. It recognizes that there are multiple technologies, and developers should be able to pick the *right tool for the job* to cover all possible use cases for enterprise user interfaces.

This book will explain why enterprise software and its user interfaces have different needs than consumer-grade software. It will also introduce you to all the current technologies that SAP offers to implement SAP Fiori–conform web user interfaces and help you to build your first project with them. Then, this book tries crystal ball gazing and looks at up-and-coming industry trends and how they relate to frontend projects driven by SAP. Please don't see this book as a comprehensive guide to teach you everything about a specific technology. Neither will it replace detailed step-by-step instructions to build, deploy, and maintain applications in a complex system landscape. Instead, it tries to teach everything that's needed to begin your own journey through the SAP UI universe.

Consumer-Grade vs. Enterprise Readiness

As mentioned in the previous section, some users are not satisfied with the user interface of the enterprise software they use daily. We believe this discrepancy originates from an unconscious comparison with consumer software and apps we interact with more often than with enterprise software. This section will focus on the fundamental differences between enterprise and consumer software to better understand the situation. It's important to understand that this is not an excuse for enterprise software to have poor user experience, but it helps to understand the developments of the past and make better decisions for the future.

Consumer-grade software is software that is designed for use by individual consumers. It is typically designed for personal use on numerous devices to reach their consumers where they are. Consumer-grade software is either sold directly to individuals through retail channels or online marketplaces or, what we see more regularly today, offered for free in exchange for the right to collect user data. For many applications, the requirements are less complex and feature-rich, and it's fine if the data lives in silos. There is usually no overarching need to integrate data from various applications. Because of these rather impulsive decisions that can be driven by recommendations of friends or social trends, there is no need for an in-depth consulting and service-level agreement (SLA), for example, for guaranteed uptime or availability of features. End users are often influenced by their emotions when they decide for one and against the other product. And as a result, the companies heavily compete for the users' preferences on a product level. The probably most difficult challenge for this kind of software is the competitive landscape it lives in. The technology companies that produce these apps frequently follow a lean methodology[1] to adapt to the users' needs. It is advantageous to be the first to release a popular feature (e.g., having a "story feature" in a social media app), having an amazingly fast loading time and a playful and intuitive user interface that makes a manual redundant. On the other hand, consumer apps usually don't suffer much if they abandon features used only by a small subset of users or change an existing feature to their advantage. They can drop support for outdated browsers or remove a chronological news feed in favor of an algorithmic one that benefits their advertisers.

But consumer apps also differentiate on a technical level. Consumer tech companies usually neither sell the underlying technology nor are they bound to strict SLAs. This makes it easier to replace technologies if it is beneficial to them. Startups and teams in larger organizations that follow

---

[1] https://theleanstartup.com/

a lean methodology have an easier time picking the latest technology and moving forward. Even if the picked technology is not the right one to address the entire market, it is often enough to test what the market desires and whether the product can be successful. This is also why so many apps are available in the Apple App Store before they become available in the Google Play Store. It's easier to develop just one app and ensure the product works. When these companies build a working product, they further improve it. This typically means speeding up the performance and optimizing the website or app for search engines. This nurtures innovation because they might need to leverage even more new technologies, such as edge computing. If they run into scaling problems along the way and find other technologies that better serve their needs, they typically switch technologies, even if it means they need to reimplement their product. This happened, among others, to Twitter and LinkedIn[2] in their early days when they had to swap the underlying technology for some parts of their backend.

Don't misunderstand this. Even for these companies, a reimplementation and technology swap is a big risk. But they don't have much choice other than to do it anyway. In general, the popularity of technologies tends to rise and fall as new technologies emerge and become more widely adopted. One example is the cross-device implementation of mobile apps. In 2009, Apache Cordova, formerly known as PhoneGap, was a very popular way to run web applications "natively" on mobile devices and access their sensors. Today, this approach feels quite outdated, and there are other popular tools like React Native, from Meta platforms, and Flutter, from Google, trying to solve the same problem. While Cordova applications were essentially web applications wrapped in a native shell, React Native and Flutter are designed to be faster and more efficient and provide a better

---

[2] www.quora.com/Why-did-Twitter-and-LinkedIn-switch-from-Rails-to-Scala-Play-in-spite-of-the-fact-that-Rails-can-be-scaled

developer experience. They also have large developer communities and ecosystems, which makes it easier to find support and resources when working with these frameworks. At the same time, Project Fugu[3] tries to bring as many mobile capabilities to the Web and reduce the need for native mobile applications. So, the next wave of tools might already be waiting for their turn.

This openness to new technologies also manifests itself in the company culture. Developers usually enjoy their work more if they can use the latest technology with a great developer experience. And this keeps them motivated to work on excellent products.

In the end, new technologies only have a relatively short time window to prove themselves. And the way to do this is to convince as many developers as possible by being reliable and having exceptional developer experience. Additionally, this technology needs to integrate with the existing technology stack. So, the bugs that each software has need to be identified as soon as possible. This is easier if many developers use this technology and collaborate on it. If enough developers find a bug in open source software, someone will fix it sooner or later. If enough developers identify a missing tool in a process, someone will invent a tool sooner or later to save time. And little by little, the ecosystem around these open source projects grows.

Enterprise software, on the other hand, is software that is designed for use in a business or organizational setting. It is typically used to support a wide range of operations, business processes, and management of an organization and is often more complex and feature-rich than consumer-grade software.

When companies need new software, they usually ask themselves whether this software provides a strategic differentiator from their competitors or if it's a commodity. A new solution for human resources (HR) might improve your processes, but it doesn't have a direct impact

---

[3] www.chromium.org/teams/web-capabilities-fugu/

on improving your product lifespan or making your customers happier. So, it's a smart idea to purchase this software over developing it in-house. The data that enterprise software processes is usually the holy grail of companies and needs to be protected carefully. Therefore, enterprise software has to fulfill high-security standards to prevent accidental exposure or criminal access to this data. Auditability is also desired to keep track of all data manipulations to be able to find out if someone fraudulently changed or removed a vital line item, say, on the balance sheet. Optimization for public search engines, on the other hand, is not desired for enterprise software.

Often, this software is designed to be used by multiple users within an organization, which comes with a larger number of possible roles. Depending on the use case, it might be necessary for specific users, like CEOs, to delegate tasks to other employees that support them, like office assistants. In practice, this is typically resolved with a customizable number of roles that can be assigned to all users. Enterprise software, in general, needs to be highly customizable to meet the specific needs and requirements of an organization. It could be relatively small requirements, such as the desire to brand the entire application suite and make it follow the corporate identity of a customer. In this case, the customization of the user interface, or its color themes, might be necessary. But customization can also be larger and mean an adaptation to the local tax situation or the company-specific logistics that differentiates them from their competitors. The requested customization could even go as far as the need to implement additional features – which is then also called extension.

To scale enterprise software and allow these modifications, it is also necessary to provide the right tools, software development kits (SDKs), interfaces, and technologies to a partner ecosystem that helps with these customization requests. Otherwise, you might not be able to serve all customers, become the bottleneck, and eventually lose business. To be able to scale this partner ecosystem, it is beneficial if your technology is easy to learn and use. Enterprise software also needs to allow fine-granular configuration options for a group of users that doesn't require software engineers at all. In an ideal world, business users can save their favorite configuration of a table view that has the right column sorting applied and irrelevant columns hidden. Good enterprise software allows this and even provides additional features for power users to help them get their job done more efficiently. This can, for example, be done with the support of recurring patterns and keyboard navigation, similar to how developers work in their favorite development environment.

Keeping the entire software portfolio of an enterprise consistent is almost a Sisyphean task. Even if all enterprise applications are perfectly integrated, a company might decide to acquire or merge with one another. This usually means in the short term that the business solutions need to be integrated, and in the long term, data is migrated from one solution to another. Therefore, integration and migration capabilities from and to third-party tools are also typical requirements. One of the most common scenarios is single sign-on with a central identity management solution, such as SAP Cloud Identity Services or Microsoft Active Directory. Such a service not only makes the life of the employees easier, but it also allows the IT department to manage all users and their access to different software products more easily. But IT also has other tasks they need to handle, that is, they have to be able to install certain apps on the hardware owned by the company. This is also another requirement of enterprise software: it needs to be remotely installable and patchable on the entire spectrum of devices. In the past, it was accepted that enterprise software only ran on Microsoft's Internet Explorer. It wasn't a big deal because Windows was so dominant and everyone was working on standardized desktops in offices. Today, the landscape has become much more complex. People want to be able to work from anywhere at any time. This includes quick checks for important emails on public transport or the approval of workflow from the waiting area of the airport. From this follows that enterprise software now needs to support multiple operating systems and device types, starting from small mobile devices and going to desktop computers.

And then there are also legal requirements for enterprise software that we haven't mentioned yet. In certain industries, such as critical infrastructure, it might be necessary to fulfill additional standards like ISO/IEC 9126[4] and its successor, ISO/IEC 25010.[5] Counties might also require that software sold in certain industries is accessible by handicapped employees or localized in the local language, such as the French Toubon Law[6] permits.

You see, many things can, but must not, break. It becomes very costly if enterprise software breaks and prevents the company from making money. Therefore, it's always important to provide high SLAs and comprehensive support and maintenance services. And customers are usually willing to pay a premium fee for these services. This explains why enterprise software is generally pricier than consumer software. And, therefore, the buying process is very rational and typically doesn't happen via self-service. Both sides, the seller and the buyer, need experts who know the software and all the possible implications, benefits, and risks that come with purchasing that software. Due to this complex sales process, large enterprise software deals usually have longer sales cycles and can take several months or even years until they go live. Because of this long time, people often think features shouldn't change much in consecutive releases, as this might change the agreed-upon scope of the software and related training. Overall, many factors slow down the innovation speed compared to consumer-grade software.

All the abovementioned factors contribute to a certain degree of risk aversion. Betting on a technology that might not exist in five years is a big risk, which comes with technical debt, and might result in unpredictable financial obligations. So many software vendors try to build everything on their own, which can also be very costly. Another popular approach to

---

[4] https://en.wikipedia.org/wiki/ISO/IEC_9126
[5] https://en.wikipedia.org/wiki/ISO/IEC_9126#Developments
[6] https://en.wikipedia.org/wiki/Toubon_Law

avoid risk is purchasing using technology that has guaranteed SLAs and a predictable lifetime horizon, which is just about the opposite of open source software.

Enterprise Software – Today and in the Future

To be honest, the risk aversion and innovation angst from the previous section could cause a negative outlook on the future of enterprise software – at least from a tech-savvy perspective. Luckily, we live in a time when the lines between consumer-grade and enterprise software start to blur, and both sections fade into each other. This holds especially for Software-as-a-Service (SaaS) solutions, where the big differentiators, SLA and reliability, are now almost gone. Do you remember the last time when successful consumer-grade apps, such as Google Search, Gmail, Google Maps, Instagram, or Facebook, went down for more than a few minutes? Now think of enterprise software such as SAP S/4 HANA, Microsoft Teams, Zoom, or Slack. Would you say these enterprise applications have a better SLA than large consumer-grade applications? No, because the companies behind these applications have demonstrated that it's possible to build highly available, globally localized software that can implement complex processes and be innovative and have an intuitive user experience at the same time. On top, these applications are integrated with a larger ecosystem of apps. And to the surprise of everyone, they were able to build all of this using the *latest open source technology*.

Successfully adapting open source technology was not easy for any company, and it took a change in the mindset, especially for big corporations. The mentioned companies had to invest in their workforce and contribute back to the open source software they were using to enhance them. We also saw that these companies created their own forks of open source projects. Occasionally, they even built new tools from scratch and donated them to open source foundations. Chrome, React,

Kubernetes, Angular, Material Design, VS Code, Redis, and MongoDB are just a few examples of projects that have been open sourced in the past. It's important to note that open source software is not equal to open source software. There are still multiple licenses that might restrict whether you can use the code for commercial purposes, but this is less relevant for the discussion around the enterprise readiness of this software.

In the past decade, we've also seen that subscription-based cloud software matured from a trend to the default business model for new deals.

One reason for this change is technical: it just would have been impossible to provide such services a few decades ago. For once, the Internet wasn't fast and reliable enough. But it would have also been impossible to operate a system landscape with multiple tenants for various customers while providing the same degree of customization that has been required. So, this software had to be built as a monolith. But engineers of consumer-grade software built open source technologies, like Docker and Kubernetes, that are the foundation for microservices. Microservices are an architectural style that structures an application as a collection of small, independent services that communicate with each other using well-defined interfaces, typically over a network. These services are self-contained and loosely coupled, meaning they can be developed, tested, and deployed independently. One of the main benefits of using a microservice architecture is that it allows for greater flexibility and scalability. Because each service is independent, it can be developed, deployed, and scaled independently of the rest of the application. Some voices in the industry think microservices make software development too complicated. They say this architecture is only valuable in certain situations, for example, when the solution needs to scale to many concurrent users. But even these voices recognize the limitations of monoliths in the competitive landscape and suggest a module-based architecture instead. In contrast to microservices, which are deployed independently, modules are programming-level constructs. These constructs are packaged and encapsulated pieces of software for reuse by other software. At runtime, all code runs on the same CPU. Both approaches make modifying and maintaining the application easier over time, as changes to one microservice or module do not necessarily require changes to the others. In the end, it's all about high cohesion and low coupling of source code, which is a big risk to existing, monolithic enterprise software that needs to be broken into smaller units. Monolithic software products are not very flexible and hard to adjust to the needs of the market and can therefore lead to a loss of market share and relevance.

Besides the technical reason, there is also an economic reason that favors subscription-based software from the cloud. On-premise software is considered an investment that will be used for many years. Therefore, it is a capital expenditure (CapEx) item and must either be amortized or depreciated over its lifetime. On the other hand, software subscription fees are considered day-to-day operational expenses (OpEx), such as salaries, utilities, and rent.

CapEx and OpEx are important concepts for businesses because they can affect the company's financial performance and cash flow. Therefore, the reason for the recent success of subscription-based enterprise software over on-premise software is not a purely technical one.

Open source technology, with excellent developer experience that can be used in enterprise software, opens (pun intended) entirely new opportunities. It's a benefit when newly hired developers can use technology they already know from their previous job or university projects. This existing knowledge speeds up their learning curve, and your projects will benefit, and you can scale your engineering teams and the teams in your partner ecosystem more effectively. There are other positive side effects, like the increased group of potential applicants who can apply at your company and an increased internal motivation because developers generally enjoy using the latest technology more than old, buggy technology.

With these motivated and well-trained developers, your entire engineering department becomes more innovative and can develop faster.

This paves the way for more stable services and better user interfaces. And with a better service in general, your customers might be more willing to pay a higher premium for the software overall.

But some aspects will remain different for the foreseeable future. Whether to buy enterprise software will stay a rational and often strategic buying decision. So, the buying decision will be made by the upper management, and end users will continue to have limited or no voting power for these decisions. Similarly, companies will continue to evaluate internally whether it's worth developing in-house solutions or buying a solution that can be integrated with their system landscape. And finally, great consumer software tries to maximize the time their users spend on them. But great enterprise software will try to reduce this time to an absolute minimum to allow business users to become more efficient.

So overall, our prediction for the future of enterprise software is that its purchase will remain a complex process that requires long sales cycles and comes with consultation and support services. However, the usability and UX of this software will be measured against the simplicity, availability, and reliability of consumer-grade software. Therefore, enterprise software needs to be built with the same open source technologies that make global consumer-grade software so successful. And companies using this software need to contribute back to the open source ecosystem, to keep the development cycle going.

A Note on Development Environments

Like all good craftspeople, every developer has a preference for tools. Think of an electrician's tool set as a metaphor: while the job calls for pliers, there are still different types. A side cutter needs more work to get the shape into things, yet it allows for precision, while pincers get the job done faster but with greater strain and less precision.

It's the same for developer tooling – one person prefers a variation of pliers over another, yet both are suited for the task. A prime example is the source code editor for UI development. There's a big variety of programs out there, from vim over Notepad++ and Eclipse to Visual Studio (Code), you name it. With all of them, you can edit XML documents, TypeScript files, or CSS style sheets – as they are essentially text files, and any text-capable editor is sufficient for working with them.

For this book, we work with Visual Studio Code,[7] or short VS Code, as a development editor. It is an open source text editor that can be extended via plugins to serve as a true integrated development environment (IDE), specifically for web development. Created and maintained by Microsoft and available for all major operating systems and browsers, it has become the de facto standard editor for modern UI development.

Characteristics of an Enterprise-Grade Development Environment

What sets the development environment in an enterprise apart from your next-door startup are the requirements "the" corporate IT department imposes on the developers. It can range from a locked-down computer with minimal user rights to web proxies that restrict access to a selected few websites only, up to more than one virus scanner running at the same time. From the corporate view, all of this represents countermeasures against the fear of being hacked or becoming a victim of ransomware. In these cases, a malicious actor gains access to the internal infrastructure of a company via a vulnerable client computer located in the company network, then steals sensitive data or encrypts network shares, offering decryption for a ransom only.

Yet at this time, the question that must be allowed to ask is, how likely is the preceding threat scenario to be initiated via a *developer* machine as opposed to an average office work computer? If people from middle

---

[7]https://code.visualstudio.com/

management answer this truthfully, they have to acknowledge that the chances of an attack happening by opening malicious Microsoft Office attachments or by social engineering far outweigh those of a compromised programming tool. Why impose the same security restrictions on computers intended for (UI) development as on those for regular office work?

That's precisely why we recommend enterprises regarding SAP UI development to give developers the permissions they need. **Allow them a free choice for their work environments by minimizing technical constraints that sabotage productive development processes.**

What is the cost of an impaired development environment? It prevents developers from getting in the flow, which means reduced creativity and even less productivity. And it will foster developers' urge to cut corners and circumvent the hindrances by experiments better not conducted. Overall, an impaired development environment will drain your development force of energy and require significant self-control from them to continue working, resulting in a downward spiral of team morale. Ultimately, everybody loses.

How can that be avoided?

Production Boosters

There's no one true cookbook for this. Still, specific measures should be

- *Restrict virus scanning to machine idle times*: Only trigger scans centrally at times when client computers have been idle for a certain amount of time. Additionally, exclude specific I/O-heavy directories (such as /**/*/node_modules) from scanning altogether. Otherwise, programming routines might be severely impeded.

- *Don't restrict access to new top-level domains*: Just because stackoverflow.com[8] lives as a .com domain doesn't mean that other resources are available as .com, .org, or .net only. A prime example is the new .dev top-level domain: Google states ".dev is a secure domain for developers and technology. From tools to platforms, programming languages to blogs, .dev is a home for all the interesting things that you build." (https://get.dev)

- *Don't force-proxy all Internet-facing traffic*: In the Node.js world, which plays a large role in UI development, installing modules requires traffic to central repositories such as npmjs.com. If proxied, cache-related side effects tend to occur, preventing module installation and thus hindering development work. So instead of proxying all traffic, *only do that for heavily used sites*, like stackoverflow.com,[9] where caching really pays off.

- *Don't force "always online" at edit time*: Web-based editors tend to solve two main purposes. First, they provide a quick way to set up a development environment. Second, they live on the Web and are thus less of a liability for corporate networks. While both aspects remain true, a web-based editor imposes an "always online" paradigm on your developers, introducing a single point of failure. If Internet connectivity is lost, no developer can work anymore, and productivity grinds to a halt. Additionally, "always

---

[8] http://stackoverflow.com/
[9] http://stackoverflow.com/

17

online" prevents any mobility for your workforce. Neither Wi-Fi nor cellular reception has perfect coverage in any country. Thus, laptops are merely reduced to static workstations from a programming perspective. Instead, allow for offline development capabilities: with your tech stack working in *flight mode*, a boost in production for your programming force is imminent. Feedback loop speed increases, as the overhead for network roundtrips is removed. Thus, the development environment responds faster, gets the developer in the flow, and fosters creativity.

- *Don't shut off central development resources*: Getting work done in the programming world doesn't always happen during regular 9-to-5 hours. Programming has artistic influences and requires creativity. That muse might come to the developer at odd hours, allowing them a mighty productive coding session – which cannot happen when centrally managed resources such as servers or network access, including virtual private networks (VPNs), are shut off for maintenance. The option to further ignite that creativity spark into becoming a full-on blaze should be supported by the development environment. To accomplish this, *refrain from stopping development resources in favor of their quality assurance (QA) counterparts*. Testing usually happens at times that can be well planned for. Keep things in your testing or QA landscape running at the required hours, and stop them at off-peak times to save money and environmental strain.

But the ultimate productivity boost is to trust your developers – because a locked-down developer computer not only represents a futile attempt against external threats but also constitutes a subconscious mistrust against the developers themselves. It's like asking them, "are you really capable of properly using this machine?"

Instead of giving in to fear and mistrust, allow your developers to use their tools of choice. Developers are digital craftspeople. And only when a craftsperson is allowed to use the tools they are most comfortable with, a true productivity gain can happen. If you force tooling upon the craftsperson, only habituation will take place, with the person getting adjusted to the utensils – but a rise in productivity will not materialize. Or would you tell your hired carpenter what tools to use?

Development managers will be able to speed up the feedback loop and accelerate the path to a successful product. Development environments, with creative headroom for your developers, will foster useful experiments and allow them to get in the flow. This results in increased morale, quicker feature delivery, and little work incentive needed: your developer productivity will escalate.

Our recommendation to all developers is to accompany any management effort for a fast development environment. Support them by explaining your tool choice properly. Don't just say you "like" software tooling, but illustrate why it's helpful to you and even how your colleagues might benefit when the entire team tries using it. If you make tool choices comprehensible, chances are that you're properly supported in return by getting the best tool for the buck.

Oh, and by the way, don't be afraid of the command line. The future was and is terminal. If the most important parts of your development environment and tech stack, in general, can be accessed via the command line, there will be little friction between developing locally and doing DevOps work for Continuous Integration and Delivery. For the modern SAP UI frameworks in this book, many tasks will be done in your terminal. Embrace and use it properly!

Looking Through the Frontend Lens

Not all the requirements previously discussed apply to the frontend development. The usage of microservices or batch processing of user profiles is an example of these. But this doesn't mean enterprise frontend development is a walk in the park. The frontend world has a few unique traits.

One is the development speed, which is significantly faster than in other areas of software engineering. New frameworks suddenly pop up, while a framework that was once popular can almost vanish within a few months. Several factors contribute to this rapid pace of change. Web standards, such as HTML, CSS, and JavaScript, are constantly being updated and improved. This means that frontend developers and their tools must stay up to date with the latest standards to take advantage of new features and capabilities. As web applications become more feature-rich and interactive, the frontend codebase required to build them becomes larger and more complex. This means that frontend developers constantly need new technologies to keep up. The increasing use of mobile devices to access the Web has led to a focus on optimizing websites for smaller screens and touch-based interactions. This has driven the development of new technologies and techniques for building mobile-friendly websites. The combination of these and other factors, which create a constantly evolving landscape, requires developers to stay on their toes and continually learn new skills. Many open source frameworks that were once hyped, such as AngularJS, jQuery UI, and Backbone.js, are now forgotten, and developers replaced them in their technology stack. The competitive landscape for frontend frameworks of that age is incredibly high, and they need to justify their usage every day. At the same time, enterprise developers rely on stable software that introduces as few breaking changes between updates as possible. This leads inevitably to a trade-off that every enterprise frontend framework needs to master.

The second trait of frontend development is the huge number of devices and application domains that need to be supported. Extending an enterprise desktop application that is handled with a mouse and keyboard by hundreds of users every day could be a typical task for an enterprise frontend developer. And just one month later, the same developer might be staffed, to rapidly prototype a web application that runs on mobile devices of field technicians. And this project could be followed by a task to visualize important metrics on a big LED screen in the sales department. In the past decade, SAP tested whether a single technology could be the right tool for all these jobs. After a few experiments, they learned there's nothing wrong in having multiple frameworks, with different strengths and weaknesses that complement each other, to cover all these devices and application domains. However, keeping multiple technologies in sync comes at a cost.

Enterprise software vendors are expected to have a consistent look and feel across all solutions in their portfolio. This is already hard for medium-sized companies that only have a few products. Large companies, like SAP, have hundreds of products and services and thousands of engineers. And the overall number of engineers working at third parties in the ecosystem is much larger. Therefore, it's important to have a design system that defines how the user interface should behave. Such a design system is a collection of reusable design elements and guidelines to create user interfaces for web and mobile applications. It may include elements such as colors, typography, layout grids, buttons, forms, and other UI components. A design system helps designers create new designs more efficiently by reusing existing elements and patterns. Guidelines for accessibility, responsive design, and other best practices are also often part of one. And the developers can also benefit in several ways from using a design system:

Consistency: A design system helps ensure that all elements of the user interface are consistent in terms of look and feel, making it easier for developers to build and maintain the UI. This can also make it easier for users to understand and use the interface, as they will be familiar with the layout and design patterns used throughout the product.

Reusability: A design system includes a library of reusable components developers can use to build the UI. This saves time and effort, as developers don't have to create new components from scratch for every feature or page.

Collaboration: A design system can be used by designers and developers, making it easier for them to work together and communicate about the UI. This can help streamline the development process and improve the overall quality of the product.

Scalability: A design system allows teams to easily scale their products and add new features without sacrificing consistency or quality. This can save time and resources, as developers don't have to recreate UI elements and design patterns constantly.

Besides SAP Fiori, other famous design systems were established by leading technology companies such as Material Design, Ant Design, or Foundation.

Developed by Google, Material Design is a design system that provides guidelines for visual, motion, and interaction design across various platforms and devices. It includes a reference implementation of reusable UI components in at least one technology, as well as guidelines for layout, typography, color, and more.

Ant Design is an open source design system developed by Alibaba Group that provides a library of reusable UI components, as well as guidelines for layout, typography, and other design elements. It is widely used by developers in China and around the world to build user-friendly interfaces.

Foundation is a frontend design system, by Atlassian, that provides a library of reusable UI components, as well as guidelines for layout, typography, and other design elements. It is designed to be easy to customize and is widely used by developers to build responsive and mobile-friendly websites and applications.

You see, there are many requirements for frontend frameworks in the SAP universe. Here are the ones that are the most important to us (Figure 1-1).

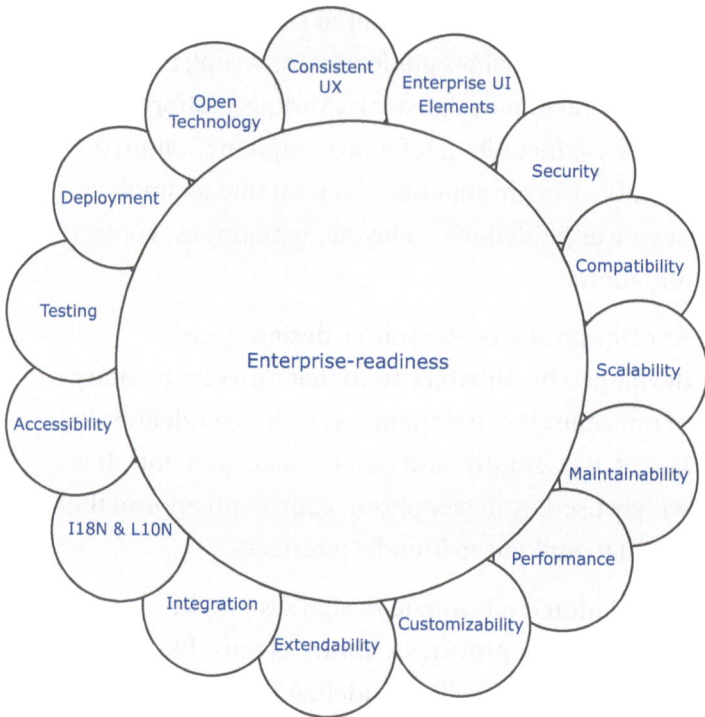

Figure 1-1. *Requirements for SAP frontend frameworks*

- *Consistent UX:* Enterprise software built with the technology needs to satisfy the expectations of its users and provide a consistent and recognizable user experience across the entire software portfolio.

- *A palette of enterprise UI elements:* It should offer a rich set of UI elements to build professional user interfaces for an enterprise context. These "UI controls" should provide keyboard navigation and build on established patterns to support power users.

- *Security:* It should have robust security features to protect against common web vulnerabilities such as cross-site scripting (XSS) and cross-site request forgery (CSRF).

- *Compatibility:* It should be compatible with a wide range of browsers and devices, including mobile devices.

- *Performance:* It should be fast and efficient, with minimal load times and resource usage.

- *Scalability:* It should be able to handle many users and a large amount of data without performance degradation.

- *Maintainability:* It should be easy to maintain and update over time while staying upgrade compatible, with good documentation and a strong community of developers.

- *Customizability:* It should allow for the creation of customized designs and layouts and should provide flexibility in terms of the types of data and functionality that can be implemented.

- *Extendability:* It should allow the extension of SAP standard apps and custom UIs at the code and application level.

- *Integration:* It should be prepared to integrate with systems and technologies from SAP and other vendors. Examples are S/4 HANA systems, SAP Business Technology Platform, SAP HANA, backend servers, databases, and third-party APIs.

–   *Globalization:* It should enable developers to create and maintain applications and websites that can be used by a global audience.

–   *Accessibility:* It should support the creation of applications and websites that are accessible to users with disabilities, like a high-contrast theme to aid visually impaired users.

–   *Testing:* It should have a robust testing infrastructure and support for automated testing to ensure the reliability and quality of the application.

–   *Deployment:* It should have good support for deployment and hosting, including options for cloud deployment and Continuous Integration/Continuous Deployment (CI/CD).

–   *Open technology:* It should ideally follow open web standards and be open source.

The SAP Fiori Design System

There is a common misconception that SAP Fiori is a set or category of products, a technology, or a framework, which is all wrong. **SAP Fiori is a design system** and SAP's answer to all the challenges mentioned in the previous section.

SAP Fiori is used by developers within and outside SAP to build consistent, high-quality user interfaces for enterprise applications. People sometimes simplify and call an app that follows the SAP Fiori design system a "(SAP) Fiori app."

SAP Fiori is designed to be flexible and scalable, allowing developers to build user interfaces that are tailored to the needs of specific users and business scenarios. The design system is also responsive, making it easy to

use on a desktop, tablet, and mobile devices (Figure 1-2). SAP Fiori follows
a set of design principles intended to guide the development of user
interfaces and ensure a consistent and intuitive user experience across
SAP applications. These principles are as follows:

Figure 1-2. *A typical SAP Fiori user interface*

1. *Role-based*: SAP Fiori apps are designed to support
   the specific tasks and responsibilities of different
   user roles within an organization.

2. *Adaptive*: SAP Fiori apps are designed to adapt to the
   needs and preferences of individual users and to work
   seamlessly across various devices and screen sizes.

3. *Coherent*: SAP Fiori apps are designed to have a consistent look and feel, with similar navigation and design elements, to provide a cohesive user experience.

4. *Simple*: SAP Fiori apps are designed to be easy to use, with a clear and intuitive user interface and minimal steps to complete tasks.

5. *Delightful*: SAP Fiori apps are designed to be visually appealing and engaging, focusing on aesthetics and usability.

Besides these design principles and the underlying design-led development process, SAP Fiori also provides guidelines for the Look, Feel, and Wording of all enterprise applications. Theming refers to the overall look and feel of the user interface, including the layout, design elements, and color scheme. This includes all the ready-to-use UI controls you can see Figure 1-2. In there, you can find labels, buttons, and tables that come out of the box with the needed features for common user operations, for example, table filtering or sorting. Each release of SAP Fiori defines a set of themes to create a cohesive and visually appealing user interface. The latest release includes a "Horizon" and a "Quartz" theme that each comes with a light and dark flavor to cover all possible user preferences. To fully support accessibility across all enterprise applications, it also comes with themes such as High-Contrast Black and High-Contrast White (Figure 1-3).

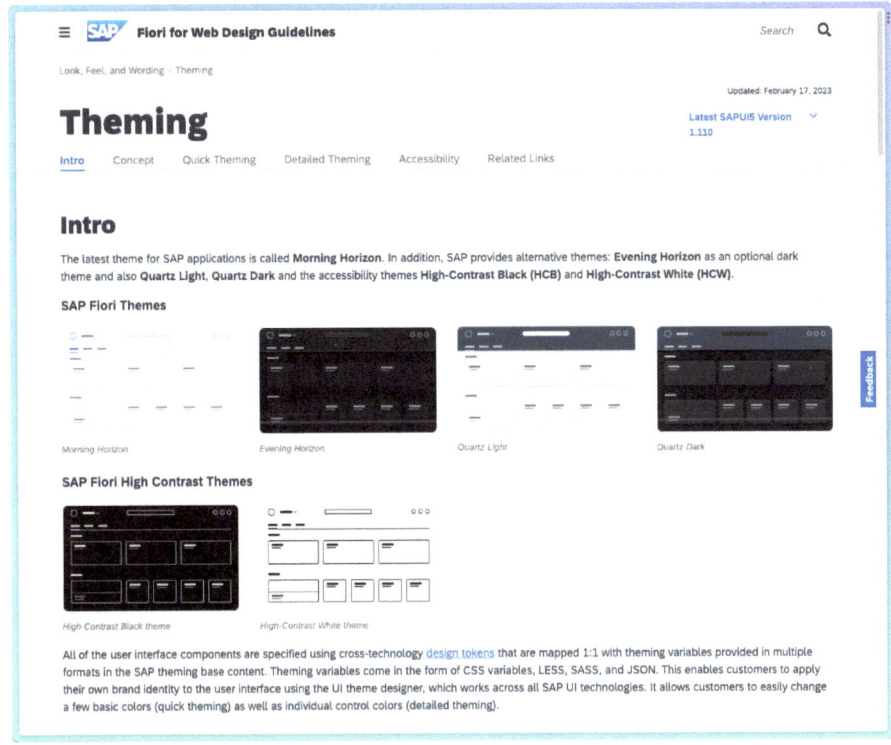

Figure 1-3. *The available SAP Fiori themes*

Iconography and typography refer to the use of icons and typing in the user interface to communicate the meaning of the UI elements. The SAP Fiori UX researchers designed two web fonts, for these two scenarios, from the ground up. Both can be downloaded from the design stencils.[10] Besides icons, SAP Fiori recently added UX illustrations as visual elements. These illustrations can be paired with written messages to better help users understand a concept, make complex ideas more relatable, and add more personality to the product by creating an emotional connection between users and the product (Figure 1-4).

---

[10] https://experience.sap.com/fiori-design-web/downloads/

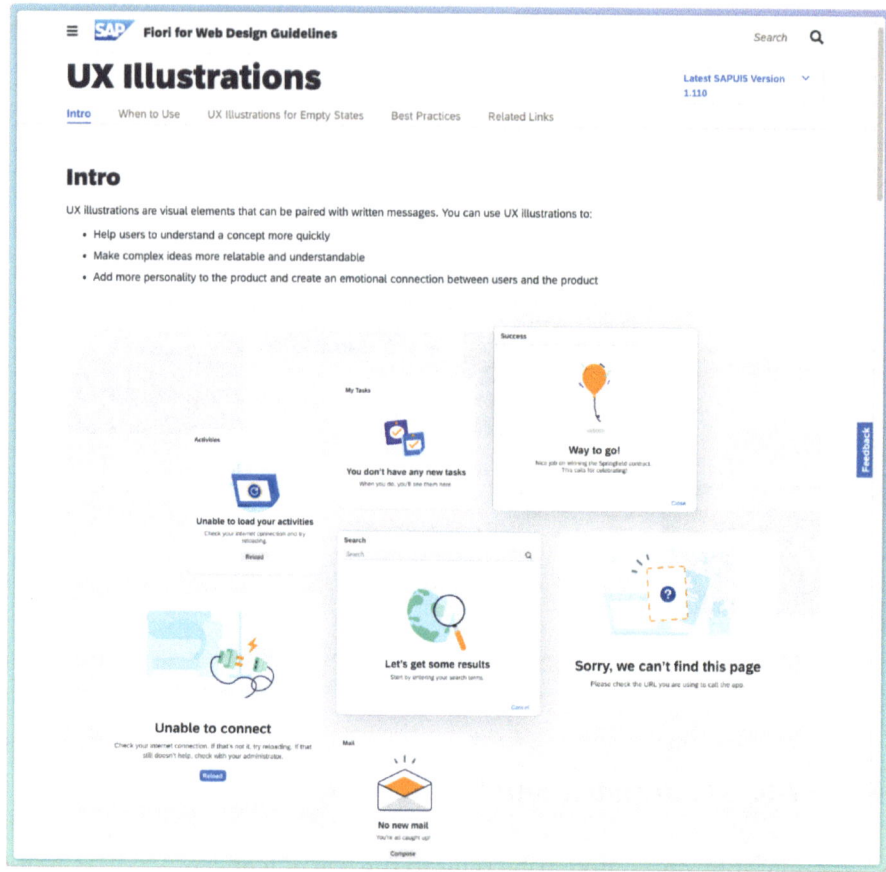

Figure 1-4. *An SAP Fiori UX illustration from the SAP UX website*

Did you know that SAP's proprietary typeface 72 was named after the year that SAP was founded?

Colors are another important element of the user interface design and can create a visual hierarchy, convey meaning, and create an emotional response in users. SAP Fiori defines a set of guidelines for the use of colors in the user interface, including the use of primary and secondary colors and the use of color to highlight critical information. Good UX guidelines also use animation and other visual effects to enhance the user experience and provide visual user feedback. This is why SAP Fiori also covers motion design guidelines for the use of subtle and purposeful animations, for example, expand and collapse animations of a panel or transitions from a list to an object page. And similarly, the design guidelines also define UI texts and their use of clear and concise language and appropriate tone and style.

All the abovementioned features of SAP Fiori are common to design systems – even the ones that are not related to enterprise software at all. So let's look at the patterns for enterprise software, where multiple users often work on the same record. Imagine a situation where two employees want to update the shipping address of one and the same order because the customer first sent an email and then decided to call the hotline. These, and similar situations, frequently happen in larger organizations, and it becomes necessary to lock business objects when someone starts to edit them. It is then helpful to display this locked state to other end users. Similarly, when a user wants to create a new object, it shall be saved for later completion, but as the object is incomplete, it shall not be stored with other completed objects. These definitions of draft and locked objects help the end users and developers. End users only need to learn this concept once and know how to deal with it across all applications, and developers don't have to invent a new concept for every business object they work on. The same goes for the clear communication of a busy state when the UI needs to wait for a long-running task in the backend. Similarly, SAP Fiori defines object handling, message handling, navigation, and other patterns (Figure 1-5).

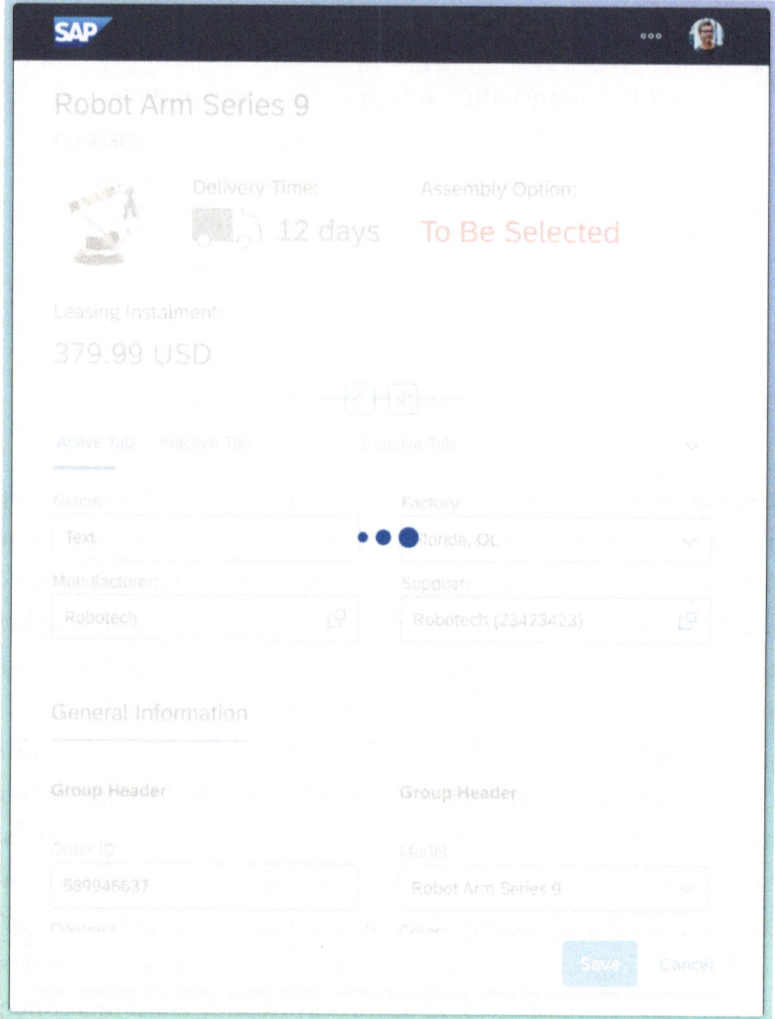

Figure 1-5. *A UI element in a busy state*

SAP Fiori builds on these patterns and defines so-called floorplans. These layout templates define the structure and organization of SAP Fiori apps. They organize content and actions within an app and are used to create a consistent and intuitive user experience across SAP applications (Figure 1-6).

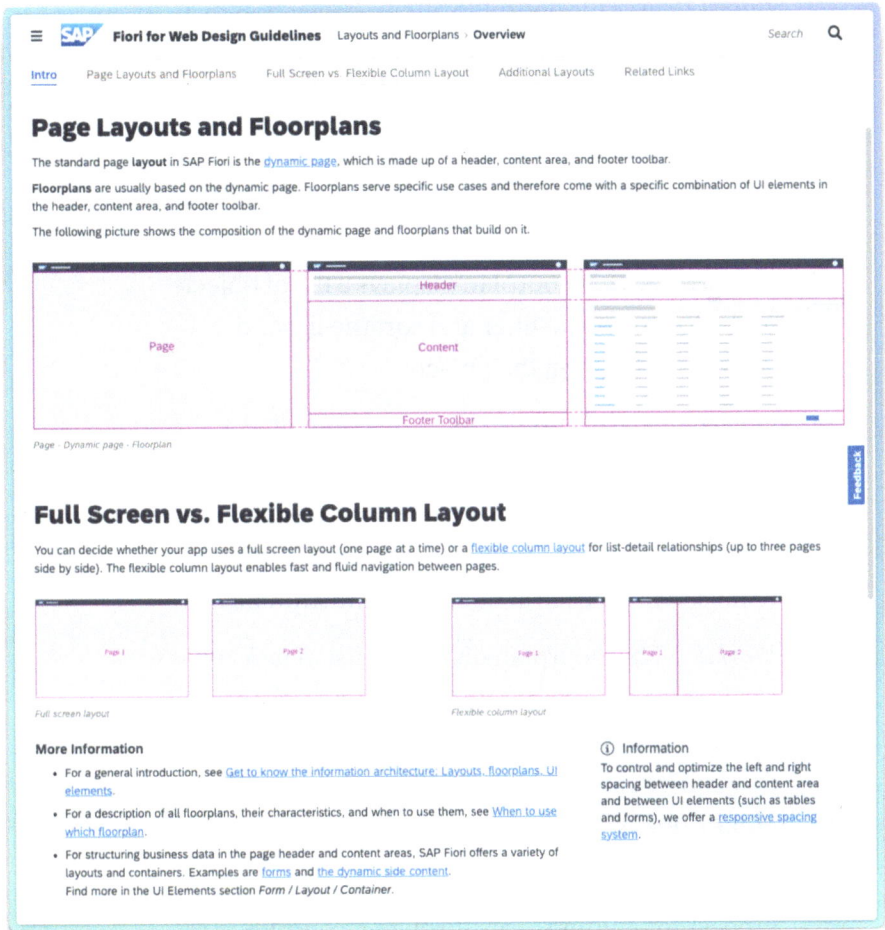

Figure 1-6. *Multiple available floorplans for pages according to the SAP UX website*

There are several SAP Fiori floorplans available, including

1. *Overview Page*: This floorplan overviews essential information and actions related to a specific business process or entity. It typically includes a list of relevant objects, such as business partners or material master records, and actions that can be performed on those objects.

2. *Initial Page*: This floorplan allows navigation to a single object, such as business partners or material master records, to view or edit it. The interaction point on the screen is a single input field that relies on assisted input to direct the user to the object in as few steps as possible.

3. *List Report*: This floorplan displays a list of objects and allows users to filter and sort the list and perform actions on the objects.

4. *Object Page*: This floorplan displays detailed information about a specific object. It typically includes a set of tabs or sections that display different information about the object, such as details, related transactions, and attachments.

5. *Worklist*: This floorplan displays a list of tasks or items that require attention or action by the user. It typically includes a set of filters and a list of items that can be marked as complete or forwarded to other users.

6. *Wizard*: This floorplan creates a step-by-step process for guiding users through a complex task or decision-making process. It typically includes a series of screens or steps that allow users to enter or select data and make decisions and provides tools for navigating between steps and reviewing or modifying data.

7. *Analytical List Page*: This floorplan displays analytical data and insights, such as key performance indicators (KPIs) and charts, and allows users to filter and drill down into the data

to explore and analyze it. It typically includes a set
of filters and a table or chart to display the data
and may also have tools for exporting or printing
the data.

An SAP Fiori app might combine various of these and less strictly
defined floorplans in a single web application that serves a specific, role-
based task. And a typical employee has to work on multiple of these apps
throughout a regular working day. This is why the SAP Fiori launchpad
(Figure 1-7) is the central access point for SAP Fiori apps. It is a hub for
launching and organizing SAP Fiori apps and is designed to provide a
consistent and intuitive user experience across SAP applications. The
SAP Fiori launchpad is organized using a set of tiles, each representing
an SAP Fiori app or a group of related apps. Tiles can be configured to
display different types of information, such as the name and icon of the
app, a summary of the app's content, or the app's status. Tiles can also be
configured to display notifications or alerts to the user.

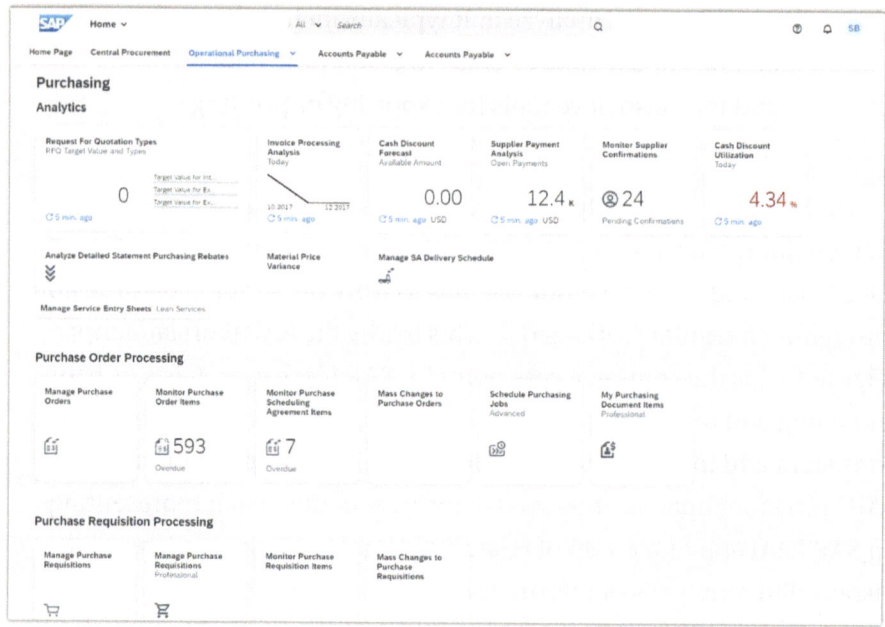

Figure 1-7. *The SAP Fiori launchpad with various tiles*

Developers and end users can customize the SAP Fiori launchpad to meet an organization's or business scenario's specific needs. This can include configuring the layout and appearance of the launchpad, adding or removing tiles, and configuring the behavior of the tiles. The SAP Fiori launchpad is a crucial element of the SAP Fiori user experience and provides a central access point for SAP Fiori apps. As a developer, you can customize the launchpad to meet the specific needs of your organization or business scenario and provide a consistent and intuitive user experience.

The full extent of the history and all aspects of SAP Fiori is written down online.[11] For completion, we briefly want to recap the history here. This design concept was started by a team of UX engineers at SAP that already had a great experience with web technologies and enterprise

---

[11] https://experience.sap.com/fiori-design-web/

software. And over the years, they refined the guideline to the concise definition we have today, and this evolution will continue. The first milestone was in 2013 when SAP drove an initiative to reimagine a mobile-first approach for the 25 most frequently used business use cases. This included tasks like creating a leave request or a sales order. SAP Fiori apps were implemented as web applications in this first version using SAPUI5 versions 1.26–1.38. SAPUI5 is SAP's strategic web UI technology: it is the reference implementation of the SAP Fiori design system to build enterprise-grade SAP Fiori applications efficiently. You will find more on SAPUI5 in the next chapter!

Once hundreds of customers confirmed the SAP Fiori design concept, it was time to scale the role-based and task-focused design approach to complex enterprise scenarios. This meant that the SAP Fiori 2.0 expanded in 2016 to more new applications and simultaneously was applied on top of existing SAP products to redesign classic UI technologies like SAP GUI for Web, WebDynpro, and CRM WebClient. With *SAP Fiori for iOS* and *SAP Fiori for Android*, two years later, SAP brought its mobile-first design approach to native mobile platforms. This company-wide adoption of the design guidelines led to multiple applications that have a similar experience but are still separated from each other. A leave request app in the HR solution looked similar to an expense report in the Travel & Expense (T&E) solution but was accessed fundamentally differently. SAP started to work on this in 2018 with SAP Fiori 3 and committed to a design system that would integrate the experience of all SAP products. The new, aligned shell bar with a product switcher was a central element toward this goal. The team of designers also established basic rules for consistency across all SAP products. SAP further broadened the number of technologies supporting the Fiori 3 design system. The design system has been extended beyond SAPUI5 and the mobile native technologies to include the latest web technologies like Web Components and frameworks like Angular, React, or Vue.

And this was the right choice, as some technologies are better suited than others for building certain types of applications. Different technologies are at different stages and therefore come with more or less of the enterprise-ready features we discussed in this chapter. This might sound like a disadvantage, but it might also be an opportunity for projects that have less strict requirements and follow a lean methodology. At the same time, some technologies provide a high level of abstraction, making it easier to build applications quickly, while others provide a lower level of abstraction, giving developers more control over the underlying implementation. **Ultimately, the choice of a web development framework depends on the specific needs and preferences of the developers and the project at hand.**

Summary

This chapter discussed the fundamental differences between enterprise and consumer software. We saw that open source technology is becoming increasingly popular in enterprise software and enables the creation of innovative, globalized software with an intuitive user experience. The chapter also recommended using Visual Studio Code for SAP UI development and provided specific measures to avoid impairing the development environment. SAP Fiori is the design system created by SAP to build user interfaces for enterprise applications, following a set of design principles that are role-based, adaptive, coherent, simple, and delightful. The design system also provides guidelines for theming, iconography, typography, color, animation, and visual effects and defines patterns for enterprise software such as draft, locked, and busy states.

The next chapters will focus on the four approaches available for web-based SAP Fiori application development: SAPUI5 and OpenUI5, SAP Fiori elements, UI5 Web Components, and Fundamental Library Styles. These chapters don't build on each other. So, it's OK to skip over them if you already know which technology you are keen to learn more about.

CHAPTER 2

Dual: SAPUI5 and OpenUI5

In this chapter, we will embark on a journey to explore the powerful frameworks SAPUI5 and OpenUI5, delving into their common origin, similarities, and how these frameworks have evolved over time.

To start, we'll guide you through configuring your development environment, installing the necessary tools, and setting up the project structure. With a properly configured workspace, you'll be equipped to dive into the captivating world of UI development using SAPUI5 and OpenUI5.

We will also take you through a hands-on tutorial on how to bootstrap your project, allowing you to build a basic user interface from scratch. As we progress, we'll explore advanced techniques for enriching the UI by incorporating additional controls and features.

No application is complete without robust testing and debugging capabilities, and we'll emphasize their importance. We'll cover unit tests, integration tests, and end-to-end tests, ensuring the reliability and stability of your application throughout its life cycle.

Before concluding the chapter, we'll shift our focus to the realm of Continuous Deployment and will reflect on the road ahead and explore the immense potential of SAPUI5 and OpenUI5 in developing enterprise applications that are scalable, feature-rich, and future-proof.

© Marius Obert and Volker Buzek 2023
M. Obert and V. Buzek, *SAP UI Frameworks for Enterprise Developers*,
https://doi.org/10.1007/978-1-4842-9535-9_2

What Is SAPUI5?

SAP's UI development toolkit for HTML5, or short SAPUI5, is a (mostly JavaScript) library used by developers to build web-based applications that have a modern and responsive UI. It provides a wide range of UI elements (a.k.a. "controls") and features that developers can use to create applications that look and feel great on various devices, including desktops, tablets, and smartphones. The project started in 2008 and evolved much since then and, over the years, went through many iterations (Figure 2-1).

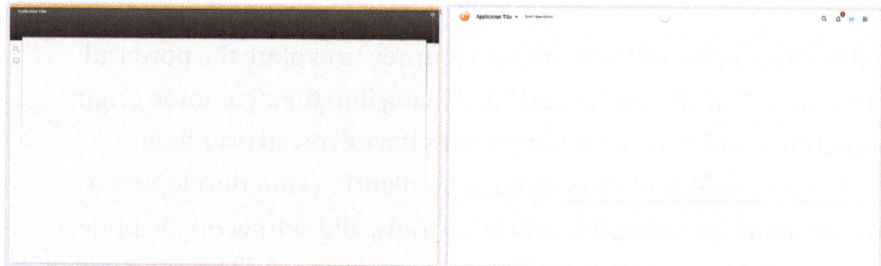

Figure 2-1. *The comparison of these two shell bar controls shows how much the framework changed over the years*

In his blog post about the history of the framework,[1] Andreas Kunz states the following goals of the initial release:

- Extensibility

  To make sure the SAPUI5 development team won't be a bottleneck for new application features.

- Allowing access to browser features

  For usage of modern browser features to compete with other web applications.

---

[1] https://blogs.sap.com/2020/11/04/a-brief-history-of-openui5-and-sapui5/

- Loose coupling with backend systems

  Allow different technology stacks, for example,
  ones used by acquiring companies that have a
  different technology stack than Advanced Business
  Application Programming (ABAP) or Java.

- Loose coupling with backend versions

  By decoupling the application and the framework,
  customers can update each individually. The
  framework should have long backward compatibility[2]
  to keep applications built years back still running.

These goals navigated the client-side rendering framework on the right
path to become SAP's standard for web-based user interfaces. And it was
the inspiration for SAP Fiori and its design system five years later in 2013.

With Backbone.js and AngularJS, there were a few popular open source
frontend frameworks around during that time, and the developers using
SAPUI5 were wishing that SAP would also open source its framework. As
SAPUI5 itself leverages many open source components, it felt only natural
to the development team to give back to the open source community. On
October 9, 2014, Peter Muessig committed the first code[3] to the new open
source sibling of SAPUI5: **OpenUI5.**

Since then and until today, this framework has been one of SAP's most
popular open source projects on GitHub. It includes the core of the SAPUI5
framework and a considerable number of control libraries, mostly artifacts
developed by teams in the UI department of SAP. Some other controls,
mostly developed in other departments, are not part of OpenUI5. They

---

[2] https://sdk.openui5.org/versionoverview.html
[3] https://github.com/SAP/openui5/commit/707bfef98f188c03bab45626ebf8
af48cddac997

remain only available in SAPUI5 because they are either too specific to SAP use cases or contain intellectual property not intended to be given away under an open source license.

Today, we often only use the term "UI5," which is not clearly defined but mostly refers to the large overlapping part of both frameworks.

Setting Up Your Workspace

An IDE provides a comprehensive set of tools for software development. It typically includes a code editor, a compiler or interpreter, a debugger, and a graphical user interface (GUI) for managing the development process. Source control management (SCM), like git,[4] is used to track changes to a codebase and collaborate with other team members.

When you install an IDE, its plugins, and all other tools on your own machine, you have complete control over the version of the software that you are using. This can be important if you need to use a specific version of the software or if you want to ensure that you are using the most up-to-date version. Installing these tools yourself also allows you to customize your development environment to meet your specific needs. You can choose the plugins and settings that suit your workflow best, rather than being limited to a preconfigured setup. Nowadays, many tools are offered for free. So having full freedom allows you to choose the tools that best fit your budget. Installing and configuring them is also a great learning opportunity. By doing this, you can gain a deeper understanding of the tools that you are using and how they work together. Equipped with this knowledge, you can also work on projects from any location, as long as you have your computer with you. This can be especially useful if you work on multiple machines or if you need to work offline.

---

[4] https://git-scm.com/

In corporate environments, with your computer managed by an IT department or system administrator, there's a constant struggle smoldering between freedom of choosing your own development tooling and centrally applied security restrictions. As addressed in detail in the opening chapter, installing an IDE and adjacent tooling can be a valuable learning opportunity. This section shall guide you to the tools that enable your first OpenUI5 project.

Git is a version control system that allows developers to track changes to their code and collaborate with other team members. We recommend installing git[5] as it allows developers to save versions of their code, revert to previous versions, and merge changes made by other team members.

Node.js is a JavaScript runtime built on Chrome's V8 JavaScript engine. It allows developers to run JavaScript on the server side, creating applications that can be run outside a web browser. Node.js is widely used for developer tooling that needs to run on multiple operating systems. To install Node.js, download the installer file from the official website,[6] run it, and follow the prompts.

By default, Node.js comes with the package manager npm. It is a package manager for the JavaScript programming language that helps developers to install and manage the packages that they need for their projects. Execute the following command in the terminal to install the needed tools:

```
npm install --global @ui5/cli yo generator-easy-ui5
```

As mentioned before, VS Code is highly customizable, allowing developers to adjust the editor to their preferred workflow, and, therefore, also our recommendation for this task.

---

[5] https://git-scm.com/downloads
[6] https://nodejs.org/en/download/

Download VS Code[7] from the official website. Once installed, add extensions from the extension marketplace (Figure 2-2).

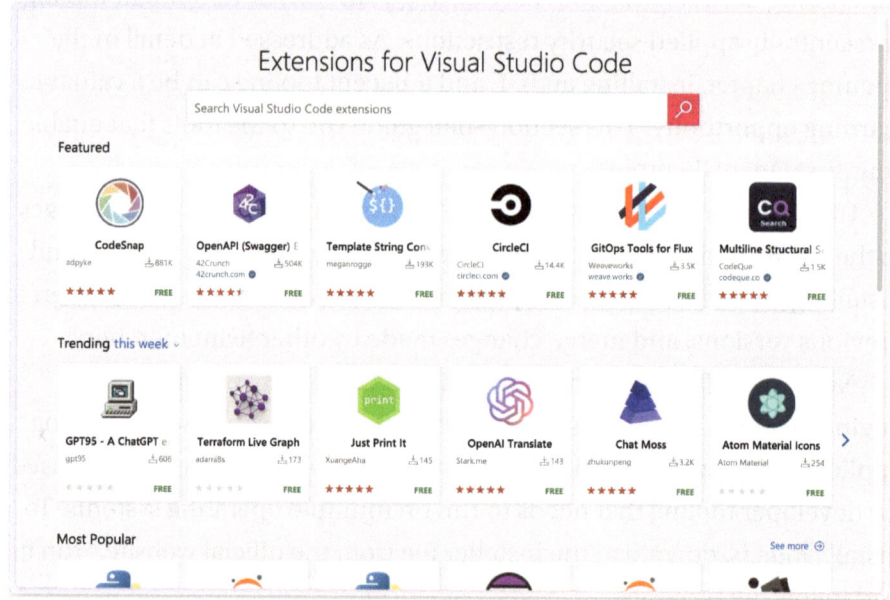

Figure 2-2. *VS Code extension marketplace*

There, you can find the following recommended extensions.

- SAP Fiori tools – Application Modeler[8]

- SAP Fiori tools – Guided Development[9]

- SAP Fiori tools – Service Modeler[10]

---

[7] https://code.visualstudio.com/download

[8] https://marketplace.visualstudio.com/items?itemName=SAPSE.
sap-ux-application-modeler-extension

[9] https://marketplace.visualstudio.com/items?itemName=SAPSE.
sap-ux-help-extension

[10] https://marketplace.visualstudio.com/items?itemName=SAPSE.
sap-ux-service-modeler-extension

- SAP Fiori tools – XML Annotation Language Server[11]

- UI5 Language Assistant[12]

- Application Wizard (UX for Yeoman generators)[13]

- XML Toolkit (Language support for XML)[14]

Learning by Doing

In this section, we will build a web app with OpenUI5. The app will be able to show a landing page that follows the best practices defined by the SAP Fiori design system and offer two features to manage beverage data:

A.   Display a searchable list of all current beers (read)

B.   Provide a form to add new beers (create)

While going through the different build steps, we'll introduce you to the most fundamental concepts of UI5 and show how to use them in a practical scenario.

Bootstrap the Project

The **easy-ui5 generator** offers numerous options to bootstrap new UI5 projects. Invoke the tool with the following command from your terminal and complete the inquiries as displayed here (Figure 2-3):

```
yo easy-ui5 project app
```

---

[11] https://marketplace.visualstudio.com/items?itemName=SAPSE.
sap-ux-annotation-modeler-extension

[12] https://marketplace.visualstudio.com/items?itemName=SAPOSS.
vscode-ui5-language-assistant

[13] https://marketplace.visualstudio.com/items?itemName=SAPOS.yeoman-ui

[14] https://marketplace.visualstudio.com/items?itemName=SAPOSS.
xml-toolkit

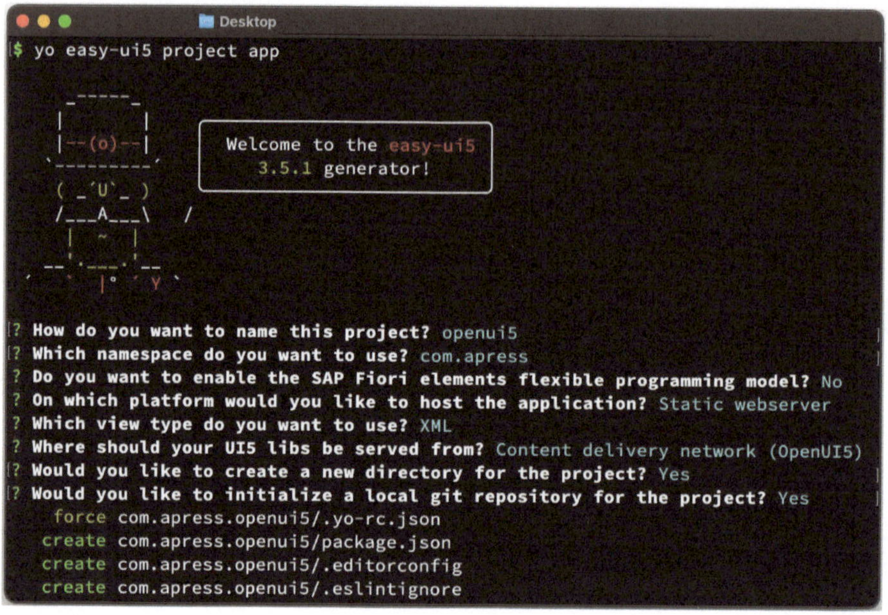

Figure 2-3. *easy-ui5 generator options*

Let's have a look at the newly created project. Go to the newly created directory and open the code with Visual Studio Code (Figure 2-4).

```
code com.apress.openui5
```

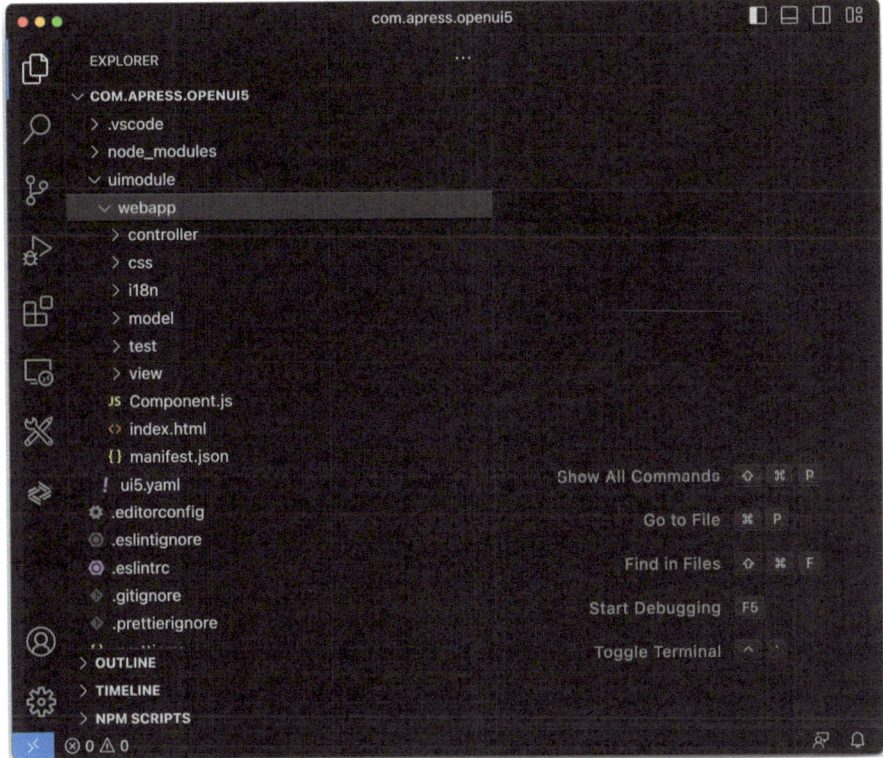

Figure 2-4. *The code for the OpenUI5 app can be found in the directory* webapp

A typical OpenUI5 project has the following structure within the webapp folder:

- controller/: This directory contains the JavaScript controllers for the application. Each controller is responsible for handling the logic for a specific view.

- css/: This directory contains the CSS styles for the application.

47

- i18n/: This directory contains the translation files for the application. These files contain the translated strings for each language that the application supports. They are usually named using the language code and country code, such as i18n_en.properties for English or i18n_de_DE.properties for German spoken in Germany.

- model/: This directory contains the models for the application. A model represents the data for the application, bound to UI controls, and communicates with the backend to fetch and update the data.

- test/: This directory contains the test files for the application. These can include unit tests, integration tests, and end-to-end tests.

- view/: This directory contains the XML views for the application. Each view corresponds to a specific page or screen in the application.

- Component.js: This file defines the root component of the application. It specifies the metadata for the component, such as the dependencies, models, and routing configuration.

- `index.html`: This is the main HTML file for the project. It loads the OpenUI5 library, thus "bootstrapping" the application.

- `manifest.json`: This file contains the configuration for the application. It specifies the metadata for the application, such as the title, description, and resources that should be loaded.

The easy-ui5 generator has already prepared all these things for you. This project leverages the UI5 Tooling,[15] configured in the `ui5.yaml` file, to start a local web server with a handy live-reload feature.[16] The great thing is, you don't have to worry about the setup at all and can run the application from the terminal via `npm start` (Figure 2-5).

```
npm start
```

---

[15] https://github.com/SAP/ui5-tooling
[16] https://github.com/petermuessig/ui5-ecosystem-showcase/tree/master/packages/ui5-middleware-livereload

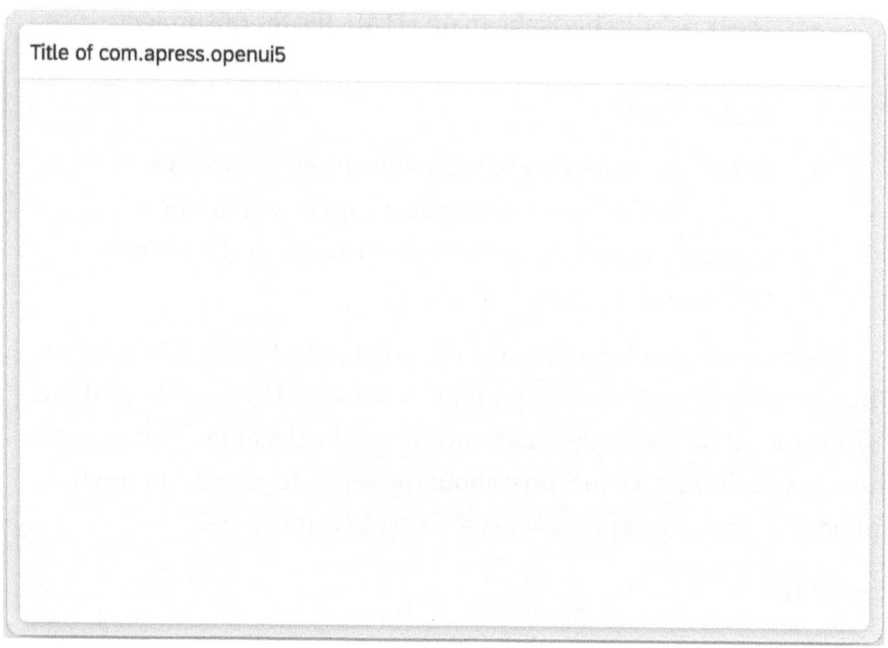

Title of com.apress.openui5

Figure 2-5. *Developer preview of a blank UI5 app*

Enrich the UI with More Controls

In UI5, a **control** is a reusable UI element that represents a specific piece of UI functionality. Controls can be simple, like a button or a label, or more complex, like a table or a form. Controls can be customized using properties, aggregations, and events.

A library (sometimes also referred to as packages or modules) is a way to organize the controls and other UI elements. Each library contains a set of related controls and UI elements and is identified by a unique name. For example, the sap.m library contains mobile controls for building responsive applications that can run on desktop and mobile devices. It includes controls for layout, navigation, input, and more. The sap.ui.layout library contains controls for layout and formatting, such as grids, forms, and responsive layout controls. It also includes utility functions for tasks such as aligning controls or calculating layout dimensions. UI5 libraries can also be created by third-party developers and made available to others.

UI5 uses the model-view-controller (MVC) design pattern to separate the data (model), presentation (view), and logic (controller) of the application. This allows for a clear separation of concerns and makes the application easier to maintain and extend.

A **view** is a container for controls and represents a specific screen or page in the application. A view is typically defined using an XML file, which specifies the layout and structure of the controls in the view. It's also possible to use JSON or JavaScript for the definition, whereas XML is the best practice. The used UI5 library can be enabled via XML Namespaces[17] xmlns. XML attributes are used to specify properties, values, and event handlers for UI controls.

For example, a view might contain a label control and a button control, like Listing 2-1.

Listing 2-1. Minimal view example

```
<mvc:View xmlns:mvc="sap.ui.core.mvc" xmlns="sap.m">
  <Label text="Hello, world!" />
  <Button text="Click me" press="onButtonPress" />
</mvc:View>
```

In this example, the text attributes specify the text that should be displayed on the button and label, respectively. The attribute is a property of the Button control, and the value specified for the attribute ("Click me") is the value that will be assigned to the property. The press attribute of the preceding example is set to onButtonPress, which specifies the name of the event handler function that should be called when the button is clicked.

---

[17] www.w3schools.com/xml/xml_namespaces.asp

To implement the event handler function, you would define it in the controller for the view. The event handler function receives an event parameter, which is an object containing information about the press event itself.

Listing 2-2 shows an example of an event handler function that could be used with the button defined earlier.

Listing 2-2. An event handler

```
onButtonPress: function(event) {
    // do something when the button is clicked
}
```

There are many different attributes that can be used in an OpenUI5 view, depending on the type of control being defined.

A **controller** is associated with a view and is responsible for handling the logic for the view. This includes handling user interactions, updating the model, and navigating to other views. The controller can access and manipulate the controls in the view through their API.

For now, let's add an appealing message to the default page that is visible when someone accesses the web app. **Modify the uimodule/ webapp/view/MainView.view.xml** file so that it looks like Listing 2-3.

Listing 2-3. Extended view

```
<mvc:View controllerName="com.apress.openui5.controller.
MainView"
    xmlns:mvc="sap.ui.core.mvc"
    xmlns:core="sap.ui.core"
    xmlns="sap.m" displayBlock="true">
    <Page title="{i18n>MainView}" id="MainView">
        <content>
```

```
    <IllustratedMessage title="{i18n>PleaseSelectAPage}"
        description="{i18n>UsePageSwitch}" />
  </content>
 </Page>
</mvc:View>
```

You might notice that we don't write the `title` and `description` directly in the view as hard-coded strings. The problem with hard-coded strings is that they make it difficult to localize the application for different languages, which is time-consuming, and they make the code more difficult to maintain. Instead of hard-coding strings in the frontend, it is better to use a translation library or resource bundle to store the strings and retrieve them at runtime. This allows the strings to be easily localized and makes the code more maintainable and reusable. We'll finish this later, but for now, the `i18n` keys are fine.

As live-reload is enabled, it should be enough to save the file and to see the change in the web app automatically (Figure 2-6).

Figure 2-6. *A UI5 app showing the illustrated image control*

Build a Reusable and Consistent Header

Let's implement a header toolbar,[18] as defined in the SAP Fiori design catalog. As our application will have multiple pages, it is a good idea for frontend developers to externalize the header into a fragment. By externalizing the header, you can reuse it across multiple pages and save a lot of time and effort, as you only have to update the header in one place instead of on every individual page. Using fragments, in general, can also help to ensure that the UI elements are consistent across all pages. This is especially important if you have a large website with many pages, as it can be difficult to manually ensure that they are consistent on every page.

---

[18] https://experience.sap.com/fiori-design-web/header-toolbar/

Externalizing fragments can also make it easier to maintain the app over time. If you need to make a change, you can do so in one place and have it reflected across the entire website.

In UI5, a fragment can contain one or more UI controls or other fragments. To use a fragment in a UI5 application, you first need to define the fragment file with XML or JavaScript. For this task, we'll focus on XML fragments only. Once the fragment is defined, it can be included in other views or fragments using the <Fragment> control. The directives for both, definition and usage, are defined by UI5 in sap.ui.core.

Listing 2-4 shows an example of a simple fragment that defines a button.

Listing 2-4. Simple XML fragment

```
<core:FragmentDefinition xmlns="sap.m" xmlns:core="sap.ui.core">
    <Button text="Click me" />
</core:FragmentDefinition>
```

To include this fragment in a view, you would use the <Fragment> control like Listing 2-5.

Listing 2-5. Include a fragment in a view

```
<core:Fragment fragmentName="my.fragment.Button" type="XML" />
```

Create a new file uimodule/webapp/view/Header.fragment.xml with this content to define a header that includes, among others, the OpenUI5 logo, a notifications counter, and a product switcher (Listing 2-6).

Listing 2-6. ShellBar header in a fragment

```
<core:FragmentDefinition xmlns="sap.f" xmlns:m="sap.m"
xmlns:core="sap.ui.core">
    <ShellBar
        title="{i18n>appTitle}"
        secondTitle="{i18n>secondTitle}"  homeIcon="https://sdk.
        openui5.org/resources/sap/ui/documentation/sdk/images/
        logo_ui5.png"
        showNotifications="true"
        showProductSwitcher="true"
        productSwitcherPressed="fnOpenSwitch"
        notificationsNumber="60"
    >
    <profile>
      <m:Avatar initials="MV" />
    </profile>
</ShellBar>
</core:FragmentDefinition>
```

This snippet makes use of the sap.f library which is not loaded by default. To avoid rendering issues, we need to add this library to the bootstrap phase of the web application. **Add the following line to the uimodule/webapp/manifest.json application descriptor** (Listing 2-7).

Listing 2-7. Library declaration in the web app descriptor

```
"sap.ui5": {
  "flexEnabled": true,
  "dependencies": {
    "minUI5Version": "1.102.0",
    "libs": {
      "sap.m": {},
```

```
  "sap.ui.core": {},
  "sap.f": {}
}
},
```

And similarly, this import also needs to be added to **uimodule/ui5.yaml**
(Listing 2-8).

Listing 2-8. Library declaration in the uimodule/ui5.yaml descriptor

```
framework:
 name: OpenUI5
 version: 1.108.0
 libraries:
   - name: sap.ui.core
   - name: sap.m
   - name: sap.f
```

A click on the ProductSwitch control shall open a popup with several
items. Each item will be represented with an icon-label combination. All
available icons can be previewed in the icon explorer.[19] This UI element
itself is also going to be a fragment. **Create a second fragment file
uimodule/webapp/view/PageSwitchPopover.fragment.xml** (Listing 2-9).

Listing 2-9. Popover in a fragment

```
<core:FragmentDefinition xmlns="sap.m" xmlns:f="sap.f"
xmlns:core="sap.ui.core">
  <ResponsivePopover placement="Bottom" showHeader="false">
     <f:ProductSwitch change="fnPageSwitch">
         <f:items>
```

---

[19] https://sdk.openui5.org/test-resources/sap/m/demokit/iconExplorer/
webapp/index.html

```
            <f:ProductSwitchItem src="sap-icon://home"
            title="{i18n>Home}" targetSrc="RouteMainView" />
            <f:ProductSwitchItem src="sap-icon://add"
            title="{i18n>AddNewBeer}" targetSrc="BeerAdd" />
            <f:ProductSwitchItem src="sap-icon://list" title=
            "{i18n>ShowAllBeers}"  targetSrc="BeerList" />
        </f:items>
    </f:ProductSwitch>
  </ResponsivePopover>
</core:FragmentDefinition>
```

Now edit the **uimodule/webapp/view/MainView.view.xml** and add the header fragment in there (Listing 2-10).

Listing 2-10. Fragment containing a header added in a view

```
<mvc:View controllerName="com.apress.openui5.controller.
MainView"
    xmlns:mvc="sap.ui.core.mvc"
    xmlns:core="sap.ui.core"
    xmlns="sap.m" displayBlock="true">
    <core:Fragment fragmentName="com.apress.openui5.view.Header"
    type="XML" />
    <Page title="{i18n>MainView}" id="MainView">
        <content>
            <IllustratedMessage title="{i18n>PleaseSelectAPage}"
            description="{i18n>UsePageSwitch}" />
        </content>
    </Page>
</mvc:View>
```

You might have noticed that the MainView includes the Header as part of its rendering process, but the PageSwitchPopover isn't included anyway at the moment. To save some time when the user accesses the pages, we only render visible UI elements. We wait and render the popover when it is accessed the first time. To accomplish this, we need to instantiate fragments programmatically. And as we want to share this behavior in all views, therefore also in all controllers, each controller needs this functionality. To make our life easier and avoid duplicate code, we move this logic to the base controller class, from which all other controller classes inherit its properties and methods.

Edit the uimodule/webapp/controller/BaseController.js file and add the highlighted lines. Make sure you don't forget to import the UI5 libraries via dependency injection in the file header (Listing 2-11).

Listing 2-11. Move shared code into the Base Controller

```
sap.ui.define(
    ["sap/ui/core/mvc/Controller", "sap/ui/core/routing/
    History", "sap/ui/core/UIComponent", "com/apress/openui5/
    model/formatter", "sap/ui/Device", "sap/ui/core/Fragment",
    "sap/m/Button", "sap/m/MessageToast"], function (Controller,
    History, UIComponent, formatter, Device, Fragment, Button,
    MessageToast) {
        "use strict";

        return Controller.extend("com.apress.openui5.controller.
        BaseController", {
            formatter: formatter,

            fnPageSwitch: function (oEvent) {
                MessageToast.show("Change event was fired from "
                + oEvent.getParameter("itemPressed").getId());
            },
```

```
fnOpenSwitch: function (oEvent) {
    const oButton = oEvent.getParameter("button");

    if (!this._pPopover) {
        this._pPopover = Fragment.load({
            id: this.getView().getId(),
            name: "com.apress.openui5.view.
            PageSwitchPopover",
            controller: this
        }).then(function (oPopover) {
            this.getView().addDependent(oPopover);
            if (Device.system.phone) {
                oPopover.setEndButton(new
                Button({ text: "{i18n>close}",
                type: "Emphasized", press: this.
                fnCloseSwitch.bind(this) }));
            }
            return oPopover;
        }.bind(this));
    }

    this._pPopover.then(function (oPopover) {
        oPopover.openBy(oButton);
    });
},
fnCloseSwitch: function () {
    this._pPopover.then(function (oPopover) {
        oPopover.close();
    });
},
```

With that change, we can interact with the header of the web app
(Figure 2-7).

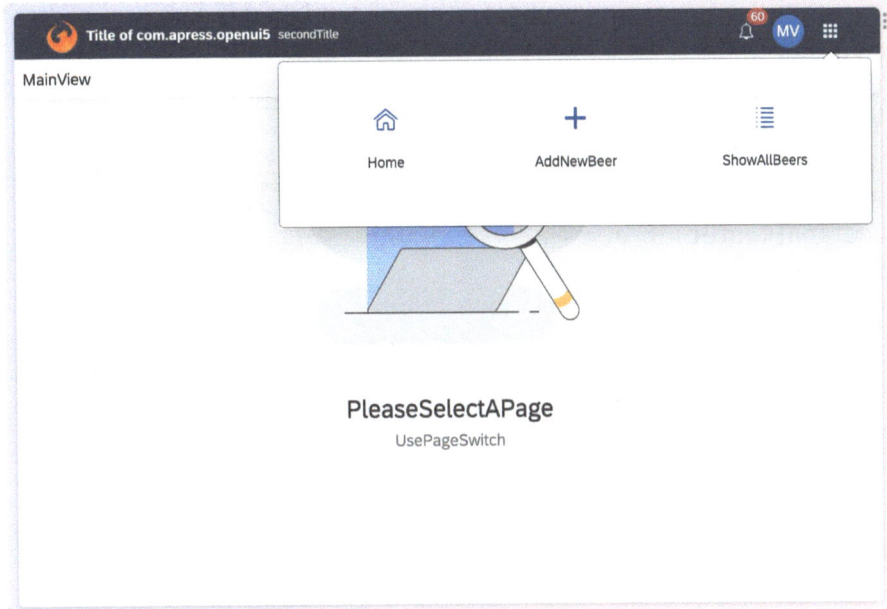

Figure 2-7. *A UI5 app showing the* shellbar *with expanded page switch*

Navigate to New Views

The router is a core component of UI5 that is responsible for handling navigation[20] in your application. It allows you to define routes, which are paths that your application can navigate to. When a user navigates to a particular route, the router will load the appropriate view and display it to the user.

The router works by listening for changes to the browser's URL. When the URL changes, it compares the new URL to the routes that you have defined in your application and determines which view to display based on the route that matches the URL.

---

[20] https://sapui5.hana.ondemand.com/sdk/#/topic/3d18f20bd2294228acb691 0d8e8a5fb5

Listing 2-12 shows an example of how you might define routes of the UI5 router.

Listing 2-12. Sample router with two routes

```
sap.ui.require(["sap/ui/core/routing/Router"],
function(Router) {
  new oRouter([{
    pattern: "",
    name: "home",
    view: "Home"
    }, {
    pattern: "products/{productId}",
    name: "product",
    view: "Product"
  }]);
});
```

In this example, we have defined two routes: one for the home page ("") and one for a product page ("products/{productId}"). The name property is used to identify the route, and the view property specifies the view that should be displayed when the route is navigated to.

In our web app, we have only one view that embeds the header toolbar, which doesn't show much reusability. Let's change this by adding two new views – one for the beer list and one for the data entry form. Each view also needs its controller, and the route needs to be defined in the manifest.json. This would usually require some boilerplate code to be set up. Luckily, the **easy-ui5** generator includes subgenerators that take that work off our shoulders, so we can focus on the important tasks.

All the terminal commands used in this section are for Unix systems like macOS and Linux. If you're using Windows, you'll need to use the Windows Subsystem for Linux to access those same commands. Windows Subsystem for Linux provides a Linux-compatible environment that you can use to run many Unix tools and applications on your Windows machine.

Open a new terminal and make sure you are in the project directory before you invoke the subgenerator (Listing 2-13).

Listing 2-13. Create a new view via `easy-ui5`

```
ls
// should print a list like this: package.json readme.md uimodule ...
yo easy-ui5 project newview
```

We need to run this command twice to generate two new views (BeerAdd and BeerList). Answer the prompts that appear when you run the generator as instructed in Table 2-1.

Table 2-1. *The options to create additional views with easy-ui5*

Prompt	First Run	Second Run
What is the name of the new view?	BeerAdd	BeerList
Would you like to create a corresponding controller as well?	Yes	Yes
Do you want to add an OPA5 page object?	No	No
Would you like to create a route in the manifest?	Yes	Yes

The generator will make all the needed modifications to the manifest. Let's edit the suggested patterns of the routes, a.k.a. the URL path used for these views. To make this adjustment, **edit the highlighted lines of the uimodule/webapp/manifest.json file** (Listing 2-14).

Listing 2-14. Route definition

```
"routes": [
      {
        "name": "RouteMainView",
        "pattern": ":?query:",
        "target": ["TargetMainView"]
      },{
        "name": "BeerAdd",
        "pattern": "beer/add",
        "target": ["TargetBeerAdd"]
      },{
        "name": "BeerList",
        "pattern": "beers",
        "target": ["TargetBeerList"]
      }]
```

For now, let's focus on the navigation between the views and keep the content of the pages as simple as possible yet differentiable. **Edit the newly generated uimodule/webapp/view/BeerList.view.xml** and add an illustrated message indicating that there is currently no list (Listing 2-15).

Listing 2-15. IllustratedMessage control in the view

```
<mvc:View controllerName="com.apress.openui5.controller.
BeerList" displayBlock="true"
 xmlns:core="sap.ui.core"
 xmlns="sap.m"
```

```
xmlns:mvc="sap.ui.core.mvc">
<core:Fragment fragmentName="com.apress.openui5.view.Header"
type="XML" />
<Page class="reducedHeight" title="{i18n>BeerList}"
id="BeerList" showNavButton="true" navButtonPress="onNavBack">
  <content>
    <IllustratedMessage illustrationType="sapIllus-
    EmptyList" />
  </content>
</Page>
</mvc:View>
```

Similarly, edit the newly generated **uimodule/webapp/view/
BeerAdd.view.xml** and add another illustrated message (Listing 2-16).

Listing 2-16. IllustratedMessage in another view

```
<mvc:View controllerName="com.apress.openui5.controller.
BeerAdd" displayBlock="true"
 xmlns:core="sap.ui.core"
 xmlns="sap.m"
 xmlns:mvc="sap.ui.core.mvc">
<core:Fragment fragmentName="com.apress.openui5.view.Header"
type="XML" />
<Page class="reducedHeight" title="{i18n>BeerAdd}"
id="BeerAdd" showNavButton="true" navButtonPress="onNavBack">
  <content>
    <IllustratedMessage illustrationType="sapIllus-
    AddPeople" />
  </content>
</Page>
</mvc:View>
```

These two views add a Page control underneath the header. But this would result in a styling issue as Page controls, by default, require 100% of the height. By creating custom CSS classes, you can override the default styles of OpenUI5 to create a unique look and feel for your app. For this reason, we made use of the classes property in the views. **Add a new style definition in the uimodule/webapp/css/style.css file** to make sure the pages don't overflow the visible viewport (Listing 2-17).

Listing 2-17. Custom CSS to adjust the page height

```css
.reducedHeight {
    height: calc(100% - 60px);
}
```

These two pages include a navigate back button in the header. Luckily, we don't need to write the onNavBack handler as this logic is already implemented in the BaseController for us. This means we only need to write a one-liner to navigate to the new pages. For this, **edit the line in the uimodule/webapp/controller/BaseController.js file** (Listing 2-18).

Listing 2-18. Implement the navigation method

```js
sap.ui.define(
    ["sap/ui/core/mvc/Controller", "sap/ui/core/routing/
    History", "sap/ui/core/UIComponent", "com/apress/openui5/
    model/formatter", "sap/ui/Device", "sap/ui/core/Fragment",
    "sap/m/Button", "sap/m/MessageToast"],
    function (Controller, History, UIComponent, formatter,
    Device, Fragment, Button, MessageToast) {
        "use strict";

        return Controller.extend("com.apress.openui5.controller.
        BaseController", {
            formatter: formatter,
```

```
        fnPageSwitch: function (oEvent) {
this.navTo(oEvent.getParameter("itemPressed").getTargetSrc());
        },
```

It's time to test the navigation now. Open the product switch control from the header and navigate to one of the new views. Notice that the URL of the web app changes when you navigate there, and it changes again when you use the back button in the page header (Figure 2-8).

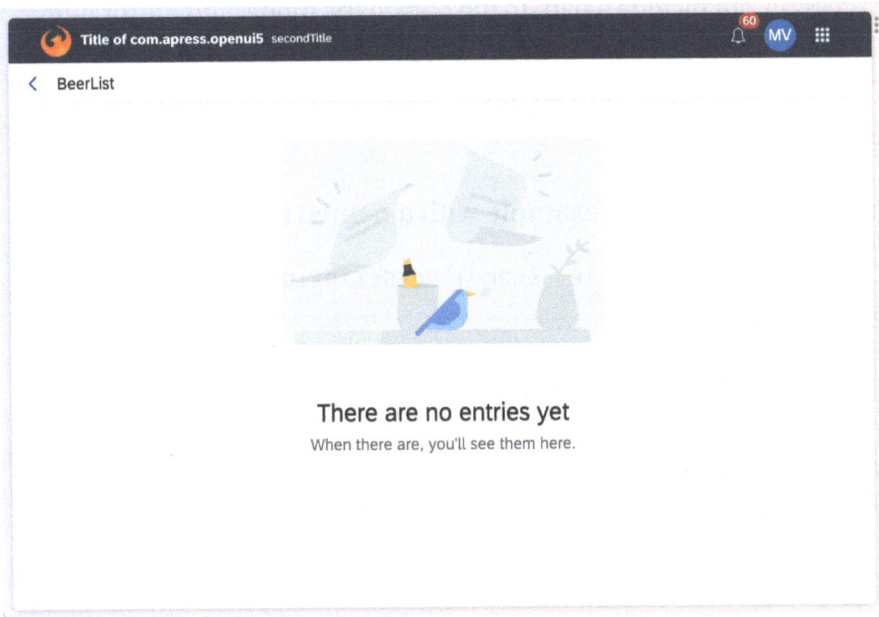

Figure 2-8. *Screenshot showing the beer list page*

Show a List of Beers

UI5 uses model bindings to bind the properties of UI controls to data stored in a model. A model is an object that represents the data in your application, and it can be bound to controls in a view using model bindings.

To create a model binding in a UI5 view, you use the { and } characters to enclose an expression that specifies the data to be bound. The expression can include a path to the data in the model and one or more binding options that control how the data is formatted or displayed.

Listing 2-19 shows an example of a simple model binding in a UI5 view.

Listing 2-19. Binding example with a named model

```
<StandardListItem title="{name}" description="{otherModel>
description}"  />
```

In this example, the `title` property of the control is bound to a data element called name in the default model. One view or control can also be bound to multiple models with different names. In this example, the property `description` is bound to the data element's `description` of the `otherModel` model. When the data in the model changes, the text displayed by the Text control will automatically be updated to reflect the new value of the name.

You can also specify binding options to control how the data is formatted or displayed. For example, you can use the **formatter** option to specify a function that should be called to format the data before it is displayed.

Listing 2-20 shows an example of a model binding with a formatter function.

Listing 2-20. A custom formatter function

```
<Text text="{path: 'price', formatter: '.formatter.
currency'}" />
```

In this example, the formatter option specifies a function called currency that will be called to format the data in the price element of the model before it is displayed by the Text control.

Expression binding is an advanced feature of UI5 that allows you to use expressions in model bindings to perform calculations or display data in a custom format. With expression binding, you can use the {= and } strings to enclose an expression that specifies how the data should be displayed or manipulated.

You can also use expression binding to access data in the model and perform calculations on it. For example, see Listing 2-21.

Listing 2-21. Expression binding calculating a price

```
<Text text="{= 'The total cost is ' + (${unitPrice} *
${quantity}) + '.' }" />
```

In this example, the unitPrice and quantity fields of the model are used to calculate the total cost, which is then displayed in the Text control. The UI5 documentation has an exhaustive list of all permitted operations[21] within expression bindings.

List bindings are another type of binding in UI5. They are used to bind an aggregatable control (such as a table or list) to a collection of data in a model. List bindings allow you to define how the data should be displayed in the control and how the control should behave when the data changes.

---

[21] https://ui5.sap.com/#/topic/daf6852a04b44d118963968a1239d2c0

To create a list binding in a UI5 view, you use the items attribute of the list control and specify a path to the collection of data in the model. You can also specify several binding options to control the behavior of the control.

Listing 2-22 shows an example of a simple list binding in an OpenUI5 view.

Listing 2-22. A minimal list binding example

```
<List items="{path: '/products'}">
  <StandardListItem title="{name}" />
</List>
```

In this example, the List control is bound to a collection of data called products in the model. The StandardListItem control is used to display each item in the collection, and the title property of the item is bound to the name element of the model.

You can also specify binding options to control the behavior of the control. For example, you can use the sorter option to specify a function that should be used to sort the data or the filters option to specify a filter that should be applied to the data.

All of this starts with a model from which there are multiple types: OData V2, OData V4, XML, Resource, and JSON models. As JSON models are the most trivial ones, let's pick this one for our sample app. Application-wide models, which are available throughout the entire application, need to be defined in the manifest.json and depend on a data source that is also defined there. Our model shall store all beverage data and be based on a .json file that we store in the model/ directory.

Modify the highlighted lines of uimodule/webapp/manifest.json file to add a new data source based on a file and define a globally scoped model based on this data source (Listing 2-23).

Listing 2-23. Global model in `manifest.json`

```
{
  "_version": "1.42.0",
  "sap.app": {
    "id": "com.apress.openui5",
    "type": "application",
    ...
    "dataSources": {
      "sampleData": {
        "uri": "model/sample.json",
        "type": "JSON"
      }
    }
  },
  "sap.ui5": {
    ...
    "models": {
      ...
      "sample": {
        "type": "sap.ui.model.json.JSONModel",
        "dataSource": "sampleData"
      }
    },
```

Also, download the sample data to **a new uimodule/webapp/model/ sample.json file**. You are welcome to add more beverages, with the same data structure, to display more items in your list (Listing 2-24).

Listing 2-24. Download sample data from the main repository

```
curl https://raw.githubusercontent.com/Apress/SAP-UI-
Frameworks-for-Enterprise-Developers-by-Marius-Obert-Volker-
Buzek/main/apps/openui5/uimodule/webapp/model/sample.json >
uimodule/webapp/model/sample.json
```

Now it's time to enrich the list view. We want to keep the
IllustratedMessage but only display it when there is no model or the list
is empty, which means it has zero list items. Both requirements can neatly
be covered when we use expression binding for the visible property.
Secondly, we'll use list binding to render one list item per beverage in
our sample data source. The title of a list item shall be the name which
is displayed above the brewery name and the concatenation of the
Alcohol by Volume (ABV) and International Bitterness Unit (IBU) values
of each beer. Finally, let's finish this view by adding a search field to the
page header.

All of this can be implemented **by adding the highlighted lines to the
uimodule/webapp/view/BeerList.view.xml** file (Listing 2-25).

Listing 2-25. View enriched by list and expression binding

```
<mvc:View controllerName="com.apress.openui5.controller.
BeerList" displayBlock="true"
    xmlns:core="sap.ui.core"
    xmlns="sap.m"
    xmlns:mvc="sap.ui.core.mvc">
    <core:Fragment fragmentName="com.apress.openui5.view.Header"
    type="XML" />
    <Page class="reducedHeight" title="{i18n>BeerList}"
    id="BeerList" showNavButton="true"
    navButtonPress="onNavBack">
        <subHeader>
            <OverflowToolbar>
```

```xml
        <SearchField liveChange=".onSearch"
        width="100%" />
    </OverflowToolbar>
</subHeader>
<content>
    <IllustratedMessage illustrationType="sapIllus-
    EmptyList" visible="{= !${sample>/beers} ||
    ${sample>/beers/length} === 0  }"/>
    <List id="beerList" visible="{= ${sample>/beers/
    length} > 0 }" items="{
path: 'sample>/beers',
sorter: {
  path: 'name'
}
}">
        <items>
            <StandardListItem title="{sample>name}"
            info="{sample>brewery}" icon="{sample>logo}"
            description="{= 'ABV ' +${sample>abv} + '
            IBU ' + ${sample>ibu} }" />
        </items>
    </List>
</content>
</Page>

</mvc:View>
```

Do you see where we defined the sorting property in this code snippet?

For enabling search on the beer list, there's only one thing to do: applying the filter logic to the list binding. For this, the event handler of the search bar needs to extract the search term from the event and, if there is a search term, apply a combined filter that consists of multiple property filters to the binding. If the search term is empty, we shall display all items. To complete this, we can use undefined to apply an "empty" filter to the binding.

Implement the highlight logic in the `uimodule/webapp/controller/ BeerList.Controller.js` file (Listing 2-26).

Listing 2-26. Search enablement on list binding

```
sap.ui.define(["com/apress/openui5/controller/BaseController",
    "sap/ui/model/Filter",
    "sap/ui/model/FilterType",
    "sap/ui/model/FilterOperator"],
    function (Controller, Filter, FilterType, FilterOperator) {
        "use strict";

        return Controller.extend("com.apress.openui5.controller.
        BeerList", {

            onSearch: function (event) {
                const oList = this.byId("beerList");
                const oBinding = oList.getBinding("items");
                const query = event.getSource().getValue();

                if (query && query.length > 0) {
                    oBinding.filter(new Filter({
                        filters: [
                            new Filter("name", FilterOperator.
                            Contains, query),
                            new Filter("brewery",
                            FilterOperator.Contains, query)
                        ],
                        and: false,
                    }), FilterType.Application);
                } else {
```

```
            oBinding.filter(undefined, FilterType.
            Application);
        }
    },

  });
});
```

This page is ready now; navigate to the page and start searching the sample data file visually (Figure 2-9).

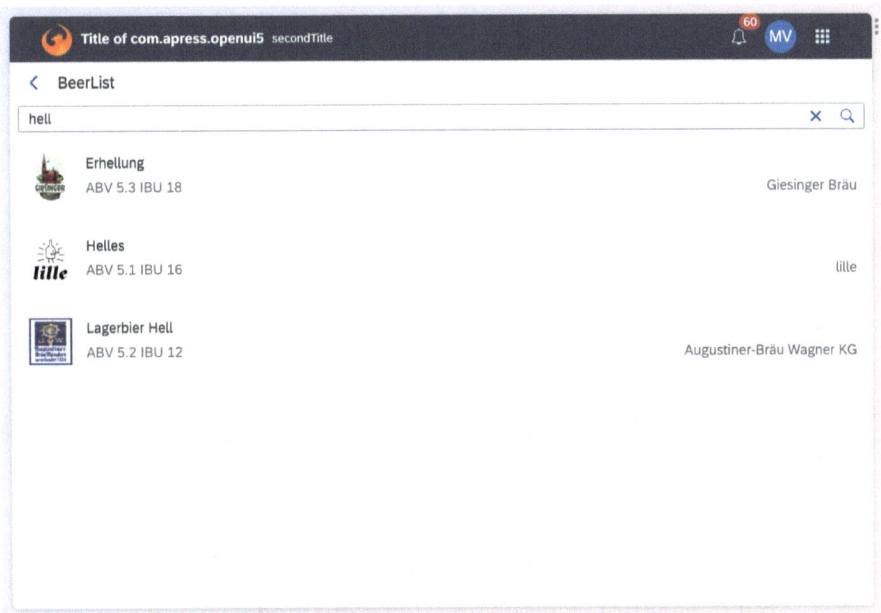

Figure 2-9. *The UI5 app with a filterable list*

Validate User Input

Input controls are UI elements that allow users to enter and edit data. Some examples of input controls include text fields, combo boxes, and date pickers. Forms are used to group related input controls and combine them in a structured way. Forms in UI5 are created using the `sap.ui.layout.form.Form` control.

Similar to lists, forms also support model binding, which allows you to bind the values of the input controls to a data model. Model binding constraints are rules that are applied to data binding. They define the conditions under which a data binding is considered valid or invalid. For example, you might define a constraint that requires a certain field to be mandatory or that a field must be a valid email address. When a user attempts to submit a form, the model binding constraints will be checked, and the form will only be submitted if all of the constraints are satisfied. This makes it easy to manage and validate the data that is entered by the user.

There are three types of data binding in UI5:

- **One-way binding** means that the value of an input control is only updated when the data model changes. This mode is used implicitly when you use expression binding or a formatter before you render the model data on the user interface.

- **One-time binding** means the same as one-way binding, but the control won't be updated if the model changes.

- **Two-way binding** means that the value of an input control is updated both when the data model changes and when the user edits the control. For example, if you bind the value of a text field to a property in a data model using two-way binding, the text field will display the value of the property. If the user edits the text field, the change will be reflected in the data model and the text field will update to show the new value.

For our learning application, we want to show a form that allows the user to add a new beer to the list. The form uses two-way binding for various input controls to enter the name, brewery, and logos. We'll add some binding constraints so that users will get a warning if they enter invalid data. Let's assume names are only valid if they have a length of at least four characters, and a valid logo needs to contain at least seven characters and start with `https://`. These checks can be declared with the `constraints` and `type` keywords with the built-in checks on the `sap.ui.model.type.String` data type. For more complex checks, it's also possible to build custom data types[22] with their own validation and formatting logic.

We also want to be able to save the beverage data we entered when the user clicks a save button. This button will use the busy indicator while the validation and saving are going on.

Open the `view/BeerAdd.view.xml` file to add buttons to the page header and the form with the input controls (Listing 2-27).

Listing 2-27. Form with multiple input fields

```
<mvc:View controllerName="com.apress.openui5.controller.
BeerAdd" displayBlock="true"
    xmlns:core="sap.ui.core"
    xmlns="sap.m"
    xmlns:f="sap.ui.layout.form"
    xmlns:l="sap.ui.layout"
    xmlns:mvc="sap.ui.core.mvc">
        <core:Fragment fragmentName="com.apress.openui5.view.
        Header" type="XML" />
        <Page class="reducedHeight">
```

---

[22] https://sapui5.hana.ondemand.com/sdk/#/topic/07e4b920f5734fd78fdaa23 6f26236d8.html

```
<customHeader>
  <Bar>
    <contentLeft>
      <Button id="back" icon="sap-icon://nav-back"
      press="onNavBack" />
    </contentLeft>
    <contentMiddle>
      <Title id="BeerAdd" text="{i18n>BeerAdd}" />
    </contentMiddle>
    <contentRight>
      <Button id="save" text="{i18n>Save}" enabled="true"
      press="handleSavePress" />
    </contentRight>
  </Bar>
</customHeader>
<content>
  <VBox class="sapUiSmallMargin">
    <f:Form id="Form1" editable="true" validationSuccess=".
    onValidationSuccess" validationError=".
    onValidationError">
      <f:title>
        <core:Title text="{i18n>General}" />
      </f:title>
      <f:layout>
        <f:ResponsiveGridLayout labelSpanXL="3"
        labelSpanL="3" labelSpanM="3" labelSpanS="12"
        adjustLabelSpan="false" emptySpanXL="4"
        emptySpanL="4" emptySpanM="4" emptySpanS="0"
        columnsXL="1" columnsL="1" columnsM="1"
        singleContainerFullSize="false" />
      </f:layout>
```

```
<f:formContainers>
  <f:FormContainer>
    <f:formElements>
      <f:FormElement label="{i18n>Name}">
        <f:fields>
          <Input required="true" value="{path: '/
          name', type: 'sap.ui.model.type.String',
          constraints: { minLength:4 }}" id="name"
          fieldGroupIds="newBeer" />
        </f:fields>
      </f:FormElement>
      <f:FormElement label="{i18n>Brewery}">
        <f:fields>
          <Input required="true" value="{path: '/
          brewery', type: 'sap.ui.model.type.String',
          constraints: { minLength:4 }}" id="brewery"
          fieldGroupIds="newBeer" />
        </f:fields>
      </f:FormElement>
      <f:FormElement label="{i18n>IBU}">
        <f:fields>
          <Input required="true" value="{path: '/
          ibu', type: 'sap.ui.model.type.Float',
          constraints: { minimum: 1 }}" id="ibu"
          fieldGroupIds="newBeer" />
        </f:fields>
      </f:FormElement>
      <f:FormElement label="{i18n>ABV}">
        <f:fields>
```

```xml
            <Input required="true" value="{path: '/
            abv', type: 'sap.ui.model.type.Float',
            constraints: { minimum: 1, maximum: 100 }}"
            id="abv" fieldGroupIds="newBeer" />
          </f:fields>
        </f:FormElement>
        <f:FormElement label="{i18n>Logo}">
          <f:fields>
            <Input value="{path: '/logo', type: 'sap.
            ui.model.type.String', constraints: {
            startsWith:'https', minLength:7  }}"
            required="true" fieldGroupIds="newBeer" />
          </f:fields>
        </f:FormElement>
      </f:formElements>
    </f:FormContainer>
  </f:formContainers>
</f:Form>
    </VBox>
  </content>
</Page>
</mvc:View>
```

The XML view earlier makes use of a default model that we haven't
defined yet. This time, we would rather not make the model available in
the entire application, but only for this particular view. This we can do
programmatically in the onInit hook right when the controller has been
initialized. Besides this one, there are other life cycle hooks.[23]

---

[23] https://ui5.sap.com/sdk/#/topic/121b8e6337d147af9819129e428f1f75

The controller also implements the event handler for the save button. First, we check whether all required inputs have valid data. If not, we need to change the ValueState of the fields with invalid input. The state of the inputs will reset to neutral if the user enters valid data, which triggers the onValidationSuccess event handler. If all data looks good, we set the button control busy to indicate to the end user that the app is working as expected. This will be useful if the frontend needs to work with a backend server that might cause some delay. To simulate this delay, we use the setTimeout function that waits three seconds before continuing. Then, we read the data from the default model of the view and append it to the component model sample. Note that this model is reset every time the web app is reloaded. This means we won't really store the new beer data, which is fine for our demo tutorial. Additionally, we reset the data of the default model to clear the form, remove the busy flag from the button control, and navigate to the beer list which now also displays the new beverage.

You can **implement all these changes in uimodule/webapp/ controller/BeerAdd.controller.js** (Listing 2-28).

Listing 2-28. Implemented event handlers for form validation

```
sap.ui.define(["com/apress/openui5/controller/BaseController",
"sap/ui/model/json/JSONModel", "sap/ui/core/ValueState"],
    function (Controller, JSONModel, ValueState) {
        "use strict";

        return Controller.extend("com.apress.openui5.controller.
        BeerAdd", {
            onInit: function () {
                this.getView().setModel(new JSONModel({}));
            },

            handleSavePress: function (event) {
                const button = event.getSource();
```

```
            if (this.triggerValidation(true)) {
                button.setBusy(true)
                const beer = this.getView().getModel().
                getData();
                const nextId = this.getModel("sample").
                getProperty("/beers/length");
                beer.id = nextId;
                setTimeout(function () {
                    button.setBusy(false);
                    this.getView().setModel(new
                    JSONModel({}));
                    this.getModel("sample").setProperty(`/
                    beers/${nextId}`, beer);
                    this.navTo("BeerList");
                }.bind(this), 3000);
            }
        },

        onValidationSuccess: function (oEvent) {
            oEvent.getSource().
            setValueState(ValueState.None);
        },

        triggerValidation: function (updateValueState) {
            const inputs = this.getView().getControlsByField
            GroupId("newBeer").filter(function (c) {
                return c.isA("sap.m.Input")
            });
            let validationStatus = true;
            inputs.forEach((function (input) {
                const oBinding = input.getBinding("value");
                try {
```

```
                    oBinding.getType() && oBinding.
                    getType().validateValue(input.
                    getValue());
                    updateValueState && input.
                    setValueState(ValueState.None);
                } catch (oException) {
                    validationStatus = false;
                    updateValueState && input.
                    setValueState(ValueState.Error);
                }
            }))

            return validationStatus;
        },
    });
});
```

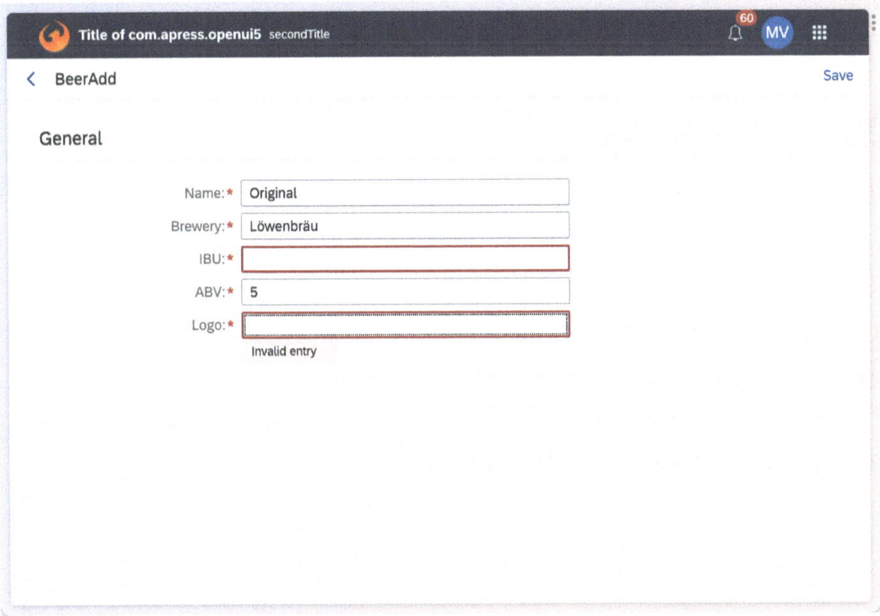

Figure 2-10. *A form to add new data*

Now go to the application and try to submit an incomplete form. You should not be able to submit the form if everything works as expected (Figure 2-10). Add valid data in the from and submit it. Once submitted, you should see a new item as shown in Figure 2-11.

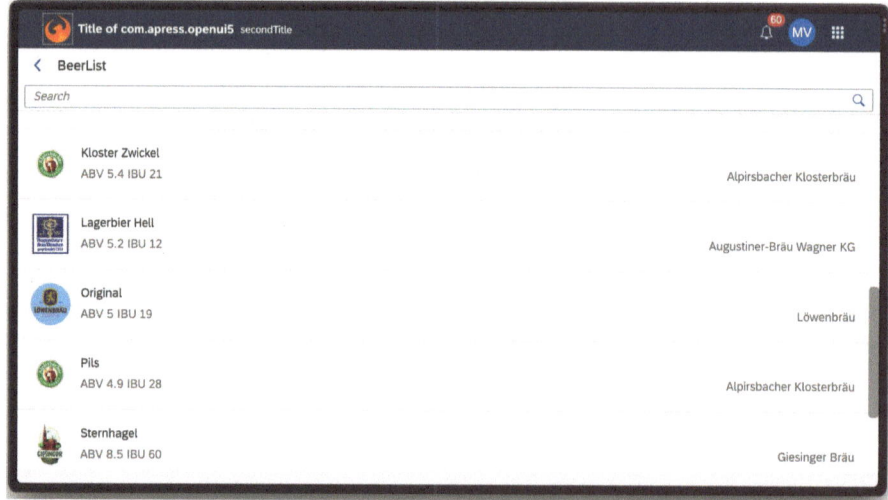

Figure 2-11. *The UI5 app after the form has been submitted*

Debug the Code

Debugging frontend apps often refers to identifying and fixing errors in the client-side code of a web application. It is an essential part of the development process as it helps to ensure that the app is functioning correctly and efficiently. Debugging tools allow developers to identify and resolve problems early in the development process.

UI5 developers use the following framework-specific debugging tools to deeper inspect running UI5 applications:

1. The Technical Information Dialog[24] displays details about the version of the framework being utilized in the app. You can enable debug mode, that is, nonminified resources, and access additional support tools to debug your app. The dialog can be opened from any UI5 app by pressing

   ```
   Ctrl Shift Left-Alt / Left-Option P
   ```
   (Windows/macOS)

2. The Diagnostics Dialog[25] is a tool that provides information about the current state of a UI5 application. It allows you to display technical information about the app and the current component hierarchy. You can even set breakpoints at framework hooks or switch the framework version being used. The dialog can be opened from any UI5 app by pressing

   ```
   Ctrl Shift Left-Alt / Left-Option S
   ```
   (Windows/macOS)

3. The UI5 Inspector[26] is a free extension for Chrome-based browsers that allows you to inspect, analyze, and support UI5-based applications. With UI5 Inspector, you can easily inspect the structure and properties of UI5 controls and view and analyze the data binding and eventing for UI5 applications (Figure 2-12).

---

[24] https://ui5.sap.com/#/topic/616a3ef07f554e20a3adf749c11f64e9.html

[25] https://ui5.sap.com/#/topic/6ec18e80b0ce47f290bc2645b0cc86e6.html

[26] https://chrome.google.com/webstore/detail/ui5-inspector/bebecogbafbi
ghhaildooiibipcnbngo?hl=en

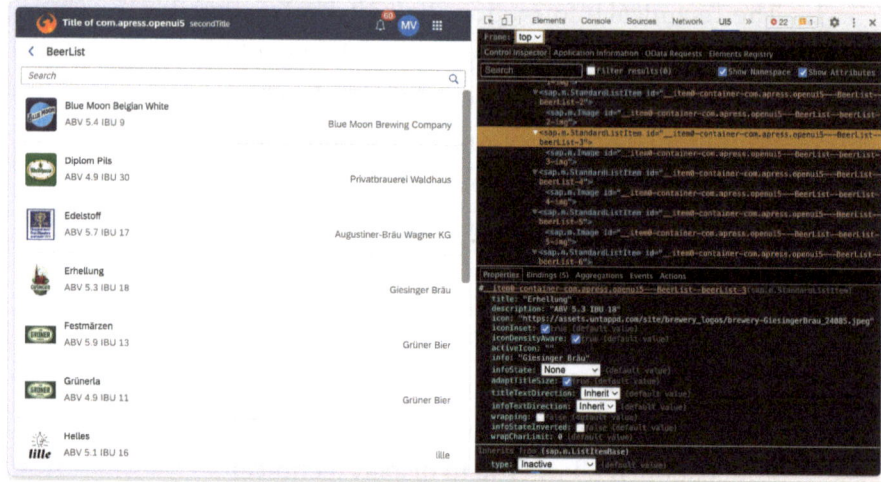

Figure 2-12. *Screenshot showing the UI5 Inspector*

Besides the framework-specific tooling, you can always leverage
a set of web authoring and debugging tools built into modern web
browsers. They allow developers to test and debug their websites and web
applications. Some common features of browser developer tools include
the following:

1. A JavaScript debugger allows developers to pause
    the execution of their code and step through it
    line by line to identify issues. A convenient trick
    to use this one is by writing the `debugger` keyword
    in JavaScript files such as a UI5 controller. Your
    browser will stop automatically when it gets to this
    directive (Figure 2-13). Once the debugger stopped,
    you can execute arbitrary JavaScript code in the
    context of this breakpoint to read properties or test
    other hypotheses.

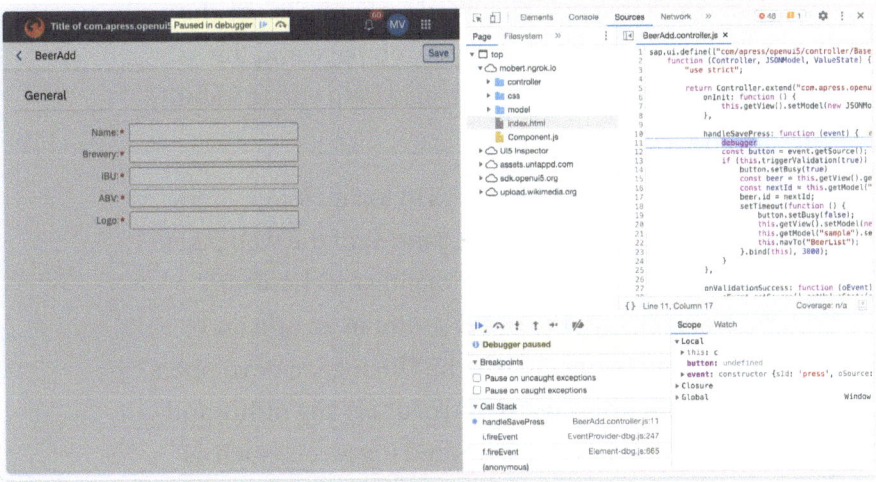

Figure 2-13. *An OpenUI5 app stopped at a breakpoint*

2. Another common requirement is that you want to
    read the properties or fire an event of a UI5 control.
    This you can also do from the debugger. The only
    thing you need is the `id` of the DOM (Document
    Object Model) element and then run the core utility
    function in Listing 2-29.

Listing 2-29. Clicking a button from the DevTools console

```
const button = sap.ui.getCore().byId("__button0");
button.firePress();
```

3. A DOM (Document Object Model) inspector allows
    developers to examine the structure and layout of a
    web page. The abovementioned UI5 Inspector can
    be seen as a UI5-specific extension of this feature.

4. A network inspector allows developers to see the
    network requests made by a web page and the
    resources it loads. This can also help you to identify
    network-related issues.

5. A console allows developers to log messages and
    run JavaScript code in the context of the web page.
    UI5 comes with a built-in logging mechanism[27]
    for tracking the flow of execution in an app and
    identifying issues. To use the UI5 logger, developers
    can require the sap/ui/core/util/Log utility and
    call the function of the desired log levels, such as
    error or info. This helps developers to categorize
    their messages and filter them as needed. Note
    that there are six different log levels in UI5, and
    they differentiate from the ones of your browser.
    To see all of them, you need to set the log level
    manually in the application code, add it via the
    URL parameter (http://localhost:8080/index.
    html?sap-ui-log-level=debug), or use the sap-
    ui-logLevel=<level> framework parameter
    (Listing 2-30).

Listing 2-30. Using UI5's logger

```
sap.ui.require(["sap/base/Log"],
  function(Log) {
    Log.error("This message will be printed by default");
    Log.info("This message will only be printed if the log
    level is set accordingly");
});
```

---

[27] https://ui5.sap.com/#/topic/9f4d62c6648a423d85aaf2bfc2c7ddfe.html

6. A source code editor allows developers to view and edit the HTML, CSS, and JavaScript of a web page. To allow debugging a browser with VS Code, a "Debug" configuration is needed. This configuration lives in the file launch.json in the folder .vscode at the root directory of your application. Google Chrome is particularly developer-friendly and thus suited to be used for debugging. Add this JSON into .vscode/launch.json (Listing 2-31).

Listing 2-31. VS Code debug configuration for Chrome

```
{
  "version": "0.2.0",
  "configurations": [
    {
      "type": "chrome",
      "request": "launch",
      "name": "Chrome for UI5 app",
      "url": "http://localhost:8080",
      "webRoot": "${workspaceFolder}/apps/openui5/
      uimodule/webapp"
    }
  ]
}
```

Specifically note the mapping of the webRoot to the UI5 application's webapp folder. This tells the Chrome/VS Code debugger where the sources of the UI5 app are located. Now, switch to VS Code's "Run and Debug" pane and hit the "play"-like looking icon in the "Chrome for UI5 app" entry in the drop-down menu. Google Chrome will open, showing the directory listing of the application's webapp folder. Set a breakpoint anywhere in a JavaScript file (e.g., in the BaseController.js).

After clicking "index.html" in the directory listing, the UI5 app will start and halt execution at the breakpoint you put in previously (Figure 2-14).

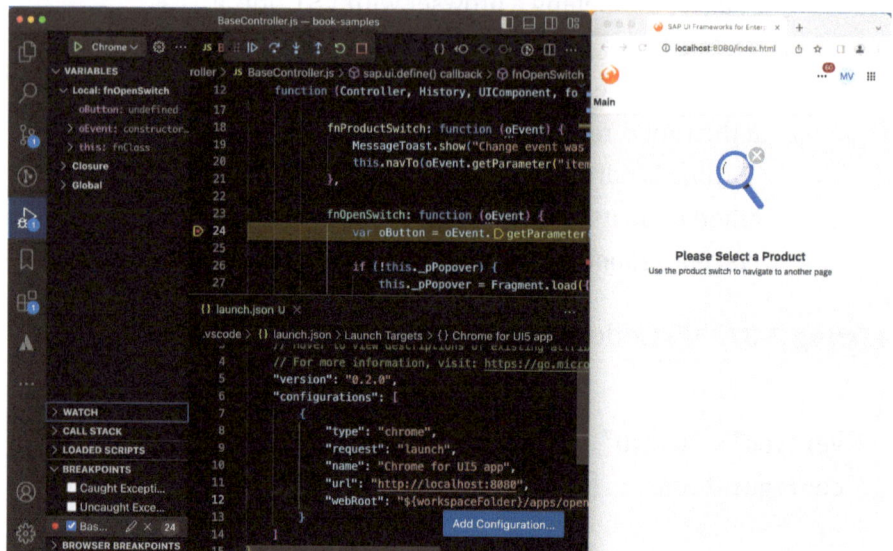

Figure 2-14. *VS Code debugger halting the browser at a breakpoint*

Translate the User Interface

In the preceding code, we already made use of the special model i18n without paying much attention to it (Listing 2-32).

Listing 2-32. i18n is applied just like data binding

```
<Label text="{i18n>labelText}" />
```

This model is called a **resource model**, which is a model that provides access to resources, such as text strings or images. These resources are typically stored in a separate file (e.g., a `.properties` file) and are accessed using the resource model. The resource model allows you to centralize your app's resources in a single location, which makes it easier to maintain and internationalize your app. Once you have a resource model, you can bind elements in your app to the resource model to display the resources in your UI. As the i18n model is used very often, it already comes preconfigured for us.

To use i18n property files in your UI5 app, you will need to create a separate file for each language or locale that you want to support. The name of each file should follow the convention i18n_LOCALE_CODE. properties, where LOCALE_CODE is the BCP-47 language tag for[28] the language or locale that the file represents (e.g., "en_US" for English, "de_DE" for German, etc.). Inside each file, you should define the resources that you want to use in your app using key-value pairs.

To fill the model with the English strings, you need to **create a new uimodule/webapp/i18n/i18n_en_US.properties** file (Listing 2-33).

Listing 2-33. The i18n file

```
# This is the resource bundle for com.apress.openui5
appTitle=SAP UI Frameworks for Enterprise Developers: A
Practical Guide
secondTitle=OpenUI5
appDescription=Description of com.apress.openui5
Home=Home
MainView=Main
AddNewBeer=Add New Beer
ShowAllBeers=Show All Beers
PleaseSelectAPage=Please Select a Page
UsePageSwitch=Use the page switch to navigate to another page
BeerAdd=Create a new Beer
BeerList=List of Beers
Save=Save
General=General
```

---

[28] www.rfc-editor.org/rfc/rfc5646.txt

```
Name=Name
Brewery=Brewery
Logo=Logo
IBU=Int. Bitterness Unit
ABV=Alcohol By Volume
```

This file will be used if the browser locale is set to "en-US." Whenever the browser locale doesn't match any of your files, it will fall back to the webapp/i18n/i18n.properties file and whatever strings are stored there. We recommend that you start populating this file with the same property keys and a translation of your choosing, that is, Listing 2-34 for German.

Listing 2-34. The German i18n file

```
# This is the resource bundle for com.apress.openui5
appTitle=SAP UI Frameworks for Enterprise Developers: A
Practical Guide
secondTitle=OpenUI5
appDescription=Beschreibung von com.apress.openui5
Home=Start
MainView=Hauptsicht
AddNewBeer=Bier Hinzufügen
ShowAllBeers=Zeige Alle Biere
PleaseSelectAPage=Wähle Eine Seite
UsePageSwitch=Benutze den Page Switch um zu einer anderen Seite
zu navigieren
BeerAdd=Neues Bier
BeerList=Bierliste
Save=Speichern
General=Eigenschaften
Name=Name
```

```
Brewery=Brauerei
Logo=Logo
IBU=Bittereinheit
ABV=Alkoholgehalt
```

Then, change the language of your browser (Figure 2-15) and refresh the application. Can you see the new strings now (Figure 2-16)?

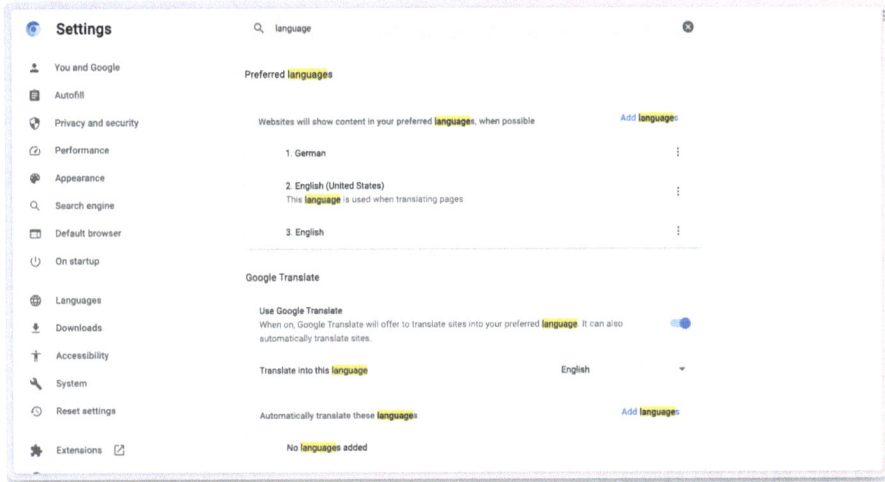

Figure 2-15. *Screenshot showing how to change the language of the browser*

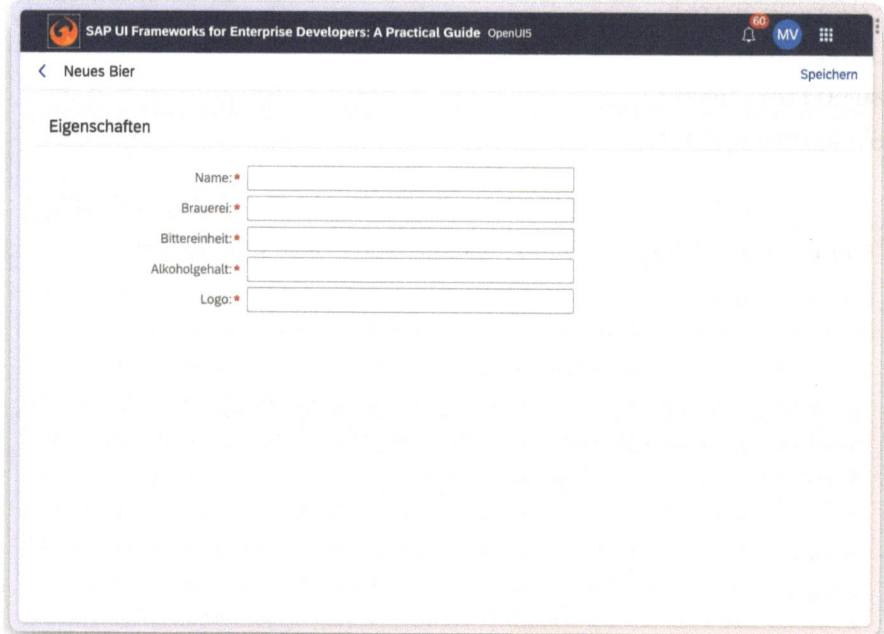

Figure 2-16. *The UI5 app with localized strings*

Build the Project

As a preparation for deploying the UI5 application, it needs to be built first. Given UI5 apps can be written in JavaScript and TypeScript, this "build" step is not a traditional "compilation," but a transpilation toward a target deployment environment. With TypeScript involved, the sources first need to be converted to JavaScript. Then, source code optimizations such as minification or the creation of the famous `Component-preload.js` file are to be conducted.

Thankfully, the UI5 Tooling[29] and its ecosystem[30] help with all of this. Install the tooling via `npm i -g @ui5/cli` so it is available to you globally, independent of the folder you work in. The UI5 Tooling contains a dedicated builder (Listing 2-35) that prepares and optimizes source code for distribution.

---

[29] `https://sap.github.io/ui5-tooling/v3/`
[30] `https://github.com/ui5-community/ui5-ecosystem-showcase`

Listing 2-35. Install and usage of the ui5 builder

```
npm i -g @ui5/cli
ui5 build --help
```

ui5 build in the root directory of the UI5 app will transpile and optimize the application's sources. The result can be used for deployment on SAP on-premise systems and the SAP Business Technology Platform (BTP).

ui5 build self-contained --all includes ui5 build and additionally packs up all used npm modules from ui5.yaml and their dependencies. The resulting folder contains all application and UI5 sources. It can subsequently be deployed to any environment with a web server and doesn't need access to the UI5 Content Delivery Network (CDN).

In case of additional requirements to either the "regular" or "self-contained" build, various UI5 Tooling tasks are available in the open source community. They range from string substitutions[31] to zipping up the build folder[32] so the app can be deployed in the HTML5 content repository[33] of the BTP.

For our sample app, cd into */apps/openui5* and run a self-contained build suited for development purposes via npm run build:dev (Figure 2-17).

---

[31] https://github.com/ui5-community/ui5-ecosystem-showcase/tree/main/packages/ui5-tooling-stringreplace

[32] https://github.com/ui5-community/ui5-ecosystem-showcase/tree/main/packages/ui5-task-zipper

[33] https://help.sap.com/docs/btp/sap-business-technology-platform/html5-application-repository

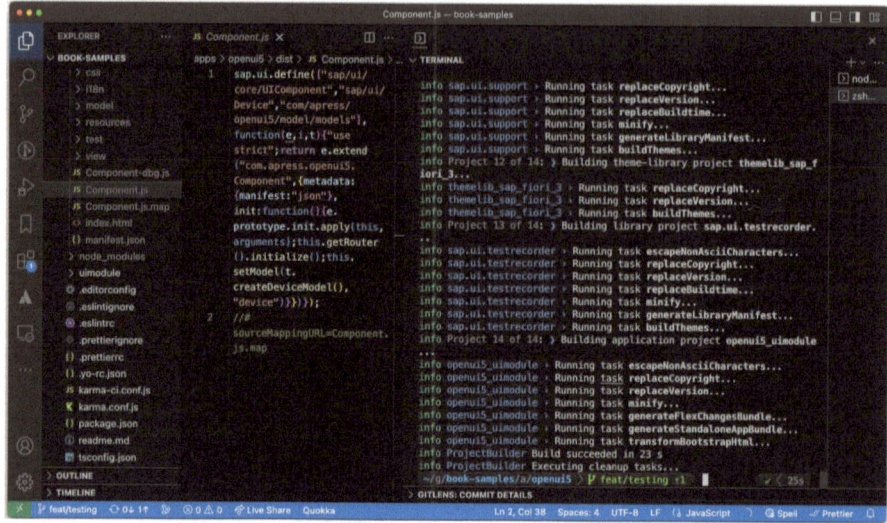

Figure 2-17. *Screenshot of the self-contained build of the sample app*

Marius wrote a popular blog post[34] on preparing a UI5 application and building it toward a target deployment scenario.

Make the App Robust

Setting up the various test scenarios requires quite some boilerplate code. Spelling it out here for you to type would not only be tedious but also waste valuable space without adding real value. Instead, we recommend that at this point, you clone the sample git repository. It contains all the scenarios set up, and you can look at both file system layout and code without being distracted by nonessential boilerplate code.

---

[34] https://blogs.sap.com/2020/10/02/serverless-sap-fiori-apps-in-sap-cloud-platform/

The technical literature is full of definitions of test scopes: "Functional," "System," "Component," "User Acceptance," and so on and so forth. In reality, it all boils down to choosing the appropriate scope in a way that has enough coverage of implemented scenarios (not lines of code!), doesn't overdo, and refrains from testing the framework itself.

Test Scopes

For this book, we cluster tests into three areas: **Unit tests** cover technical functionality, **Integration tests** refer to business processes, and **End-to-End tests** simulate user interaction. If you have lots of advanced algorithms in your app, then Unit tests are the way to assert their functionality. When there's a plethora of screen flows representing multiple business scenarios, Integration tests come into play for safeguarding the correctness of sequences. And if you rely on user interaction such as scanning a QR code, only an end-to-end test can cater toward that.

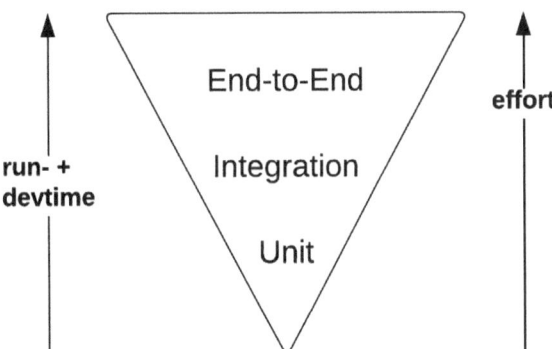

Figure 2-18. *Our UI5 testing pyramid*

At the same time, both runtime and development time increase from Unit over Integration toward the End-to-End scope (see Figure 2-18). The higher up you move in the preceding test pyramid, the longer the tests take to write and the longer they take to run! While Unit tests execute in the browser, in the same layer as UI5 itself, they tend to be very fast. An End-to-End test with wdi5 runs in Node.js, in a different layer as the browser. Communication overhead occurs, resulting in the tests to take significantly longer than a Unit test. Plan for this fact already in the work estimates!

Tests Help Realistic Work Estimates

The factor "pi" has proven to be a valid multiplier for the maximum total effort involved in a development effort that includes tests (see Figure 2-19).

If you estimate a functionality will take n hours to implement, calcuate the following:

$n*\pi$ hours for the actual implementation

$+ (n+\pi)$ hours to safeguard via Unit tests

$+ (n+2\pi)$ hours to provide Integration tests

$+ (n+3\pi)$ hours to secure via End-to-End tests

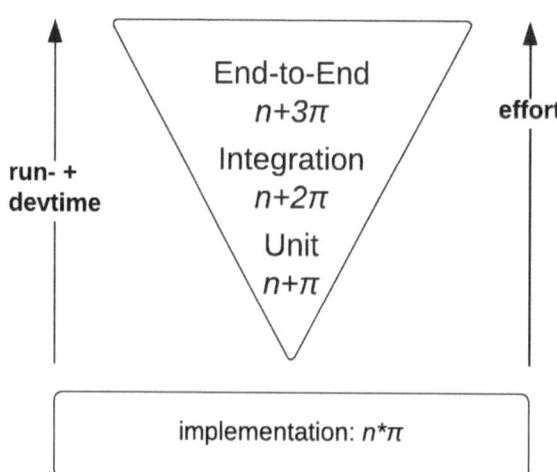

Figure 2-19. *Testing pyramid annotated with effort estimation guideline*

So if you *think* your feature will need 8 hours to implement, calculate an actual (8*3.14 =~) 24 hours.

Then expect an additional (8+3.14 =~) 11 hours for corresponding Unit tests,

plus (8 + (2*3.14) =~) 14 hours for Integration tests and
on top (8 + (3*3.14) =~) 17 hours for End-to-End tests,
resulting in a total of 66 hours.

Note that we regard this as a guide for a maximum amount of effort necessary – not all test scopes might apply, and some tests might even be doable faster than "$n + \$i*\pi$". Yet it is better to calculate more work up front than to run out of time in the doing.

Also, be aware that writing tests of any scope will most likely have you come back to the feature implementation itself and change it, even more often than not. It is just in the nature of things that once you write tests to prove certain functionalities, shortcomings, limitations, and bugs surface in their implementation. And those of course need correction, so expect a **back and forth between feature realization and verification via tests**. So writing tests in the appropriate scope is actually a part of the feature implementation! That is why **any estimates for certain functionalities should always hold the total amount of implementation and testing effort** – and not separate between the two in any planning. Following our example earlier, we'd voice 66 hours for the new feature – and not 24 for implementation and 42 for writing tests.

Tests Make You a Better and More Relaxed Developer

From a developer perspective, writing tests has two major benefits: first, it will make you more confident with your coding. Let's say you add features to your app; or you decide to refactor some implementation; or it's Saturday night and you need to provide a production-critical hotfix – when you have tests in place proving that the rest of the functionality of your app still works as designed, you can confidently make your changes. With the tests, there's a measurable quality at hand that your existing code functions as intended. When the test results remain "green," indicating expected results, the code changes you made don't introduce any regressions, a.k.a. new bugs. That builds up your dev stamina! And when you have added tests for your latest changes as well, then your confidence will even more solidify, as you put in early safeguards for the next development changes.

This leads back to the paragraph intro: the other major benefit of writing tests is that you can sleep better at night. No more brain churning around any code you put into production that might or might not or might only break things a little. Turn over in your sheets, let out that sigh of relief, and rest assured that because you wrote tests proving the correctness of your implementation, this night will go undisturbed (of code stuff, at least).

There is glory in prevention – most of all for you as a developer. Writing tests is not trivial, but worth every minute you invest into it.

Let's examine of how in UI5-land, Unit, Integration, and End-to-End tests look like.

Unit Tests

Unit tests for UI5 applications are written with QUnit2.[35] Dating back to the days of jQuery regency, you could consider QUnit old and outdated. Yet an equally valid opinion is to consider it mature and battle-tested. And in UI5-verse, it certainly gets its job done: it runs fast and is robust as in very few false positives.

Bootstrap QUnit in a dedicated HTML file from either UI5's CDN (Content Delivery Network) or from a relative location in your project (see /apps/openui5/*uimodule/webapp/test/unit/unitTests.qunit.html* in the sample repo) (Listing 2-36).

---

[35] https://qunitjs.com/

Listing 2-36. Bootstrap QUnit for Unit tests

```
<script src="https://ui5.sap.com/resources/sap/ui/thirdparty/
qunit-2.js"></script>
<script src="https://ui5.sap.com/resources/sap/ui/qunit/qunit-
junit.js"></script>
```

Then, QUnit is available in the global scope and can be used in your Unit test files just as in any other UI5 JavaScript file (see /apps/openui5*/uimodule/webapp/test/unit/controller/BeerAdd.controller.js* in the sample repo) (Listing 2-37).

Listing 2-37. A minimal Unit test

```
sap.ui.define([], function () {
  QUnit.module("...")
  QUnit.test("...", function (assert) {
    assert.ok(true)
  })
});
```

As a best practice and in preparation for automatic execution, it is recommended that all Unit test files are aggregated in a single file (AllTests.js) ... (Listing 2-38)

Listing 2-38. Unit test aggregation file

```
sap.ui.define([
  "./controller/BeerAdd.controller",
  "...",
  "..."
], function() {});
```

... and in turn scheduled to execute in a parent unitTests.qunit.js (Listing 2-39).

Listing 2-39. Starting the Unit tests

```
QUnit.config.autostart = false;
sap.ui.getCore().attachInit(function () {
  "use strict";
  sap.ui.require([
      "com/apress/openui5/test/unit/AllTests"
    ], function () {
      QUnit.start();
    });
});
```

This produces the overall setup of wrapping UI5 Unit tests as shown in Figure 2-20.

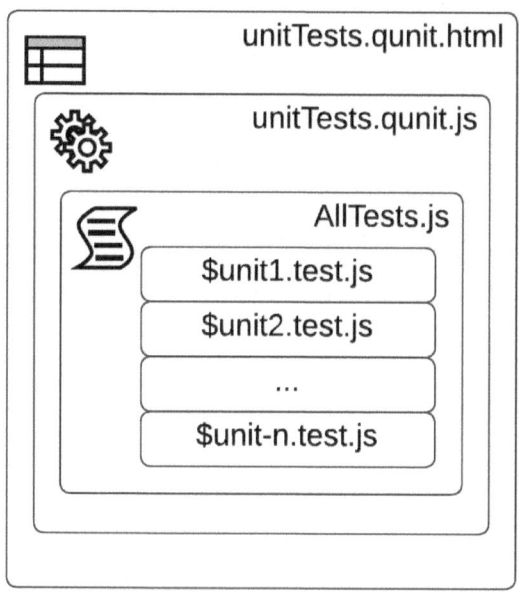

Figure 2-20. *How to organize Unit tests*

When the UI5 app was initialized via the recommended easy-ui5 generator, there's already a setup in place to reuse the preceding structure of Unit tests for automated execution. karma[36] along with the dedicated UI5 plugin karma-ui5[37] is used for that purpose. Similarly well tried and tested as QUnit2, karma lets you configure a runtime environment including a target browser to run the Unit tests in.

Look at both /apps/openui5/*karma.conf.js* and /apps/openui5/*karma-ci.conf.js* in our sample repository to get to know how to configure karma for dev time and in a Continuous Integration (CI) environment. The latter uses a "headless" browser, as in most CI environments, there is no window system active that allows a browser to start with a GUI. Instead, the browser needs to run without any UI, thus headless.

When you've cloned our repo and installed all dependencies, you have two options to run the Unit tests:

- Change into the app folder with cd apps/openui5. Then start the app via npm start, and navigate your browser to http://localhost:8080/test/unit/ unitTests.qunit.html.

- Change into the app openui5 folder with cd apps/ openui5. Then execute the tests in an automated fashion just like a CI would via npm run karma or npm run karma-ci.

By the way, there's a (somewhat aging, but still very valid) blog post series by Volker for doing "many things testing" in UI5: https://blogs. sap.com/2018/08/30/testing-ui5-apps-part-1-setup-and-unit-testing/.

---

[36] https://karma-runner.github.io/latest/index.html
[37] https://github.com/SAP/karma-ui5

Quick hints at the chapter end:

- Install type definitions for a better development experience:

  ```
  npm i –save-dev @types/sinon @types/qunit
  ```

- Add /* global QUnit, sinon */ at the top of your Unit test files for proper namespace resolution.

- Annotate first usage of QUnit and sinon with JSDoc type declaration:

  ```
  /** @type {import("qunit")} */ (QUnit).config.
  autostart = false

  /** @type {import("sinon") } */ (sinon).stub(...)
  ```

Integration Tests

When a sequence of screens and logic implement a business process, UI5 has "OPA5" as a means of testing the integration between the two. OPA5 (or "OPA" as a synonym) stands for "one-page acceptance tests for UI5" and is implemented as an additional layer on top of QUnit2 – which, as we've seen in the previous section, is used for Unit testing.

Fortunately, the basic setup is done for us by starting the project with easy-ui5. The generator produces all necessary files for Integration tests (Listing 2-40).

Listing 2-40. Integration test file system layout

```
integration
├── AllJourneys.js
├── MainJourney.js
```

```
├── arrangements
│   └── Startup.js
├── opaTests.qunit.html
├── opaTests.qunit.js
└── pages
    └── Main.js
```

A *Journey* is considered a path through the application. *Arrangements* (and *Actions* and *Assertions*) in the OPA5 context describe the behavior of the test – first, arrangements for setting up the test are done, then actions are conducted, until a certain state of the app can be asserted.

An Integration test looks similar to a Unit test (see Listing 2-37), with the specific API opaTest being used instead of QUnit.test (Listing 2-41).

Listing 2-41. A basic OPA5 Integration test

```
sap.ui.define(
  ["sap/ui/test/opaQunit"],
  /**
   * @param {typeof sap.ui.test.opaQunit} opaTest
   */
  function (opaTest) {
    QUnit.module("a Journey")
    opaTest("Should see the page", function (Given,
    When, Then) {
      // Arrangement
      Given.iStartMyApp()

      // Action
      When.onTheMainPage.ipressTheButton()

      // Assertion
      Then.onTheMainPage.iShouldSeeTheTitle()
```

```
      // Cleanup
      Then.iTeardownMyApp()
    })
  }
);
```

Generally speaking, OPA tests interact with the UI5 application on a programmatic level. There are *matchers* to locate UI5 controls and *actions* to interact with them. Both are available via the waitFor API that handles the correct order of things that otherwise are asynchronous (Listing 2-42).

Listing 2-42. OPA's matcher and action sample

```
// locating a button
When.waitFor({
  controlType: "sap.m.Button",
  matchers: new Properties({
    text: "Add New Beer"
  }),
  success: function (aButtons) {
    Opa5.assert.ok(true, "Found the button: " + aButtons[0])
  },
  errorMessage: "Did not find the button"
});

// pressing the button
When.waitFor({
  controlType: "sap.m.Button",
  matchers: new Properties({
    text: "Add New Beer"
  }),
  actions: new Press(),
  errorMessage: "couldn't press the button"
});
```

Also similar to Unit tests, the recommendation is to bootstrap OPA Integration tests in a dedicated HTML file. Then include a wrapper that aggregates all Integration tests and is capable of starting them (Figure 2-21).

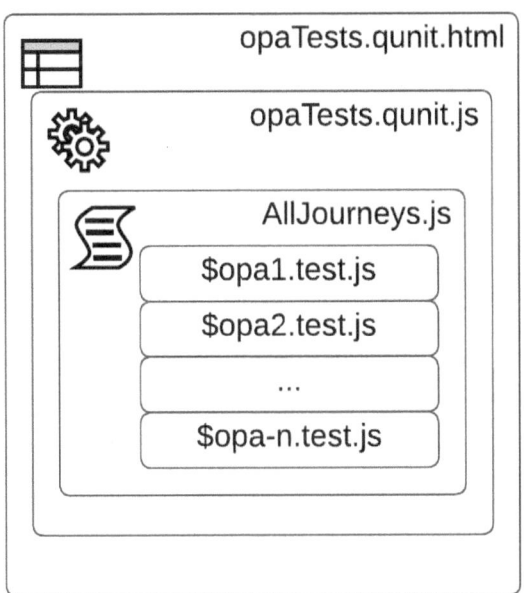

Figure 2-21. *Integration test wrapper*

It is worth mentioning that OPA tests run in sequence only – both for files and suites – meaning all *.test.js files are executed sequentially, with each QUnit.module containing opaTests also run one after the other. Specifically, the ordered run of opaTests allows for a simplified approach to writing a *Journey*. Just as coded, the steps (a.k.a. opaTest) are conducted one after another. While this adds in overall runtime, the chance of unintended side effects is reduced – the application will always be in that one state described because no additional tests are run in parallel.

Just as with Unit tests, `karma` is used for automatic execution of the Integration tests. In our sample app, it runs both the Unit and Integration tests. Look at both /apps/openui5/*karma.conf.js* and /apps/openui5/*karma-ci.conf.js* for the configuration.

Once you've cloned our repo and installed all dependencies, you have two options to run the Integration tests:

1. Change into the app folder with `cd apps/openui5`.

   Then start the app via `npm start`, and navigate your browser to `http://localhost:8080/test/integration/opaTests.qunit.html`.

2. Change into the app folder with `cd apps/openui5`.

   Then execute both Unit and Integration tests in an automated fashion just like a CI would via `npm run karma` or `npm run karma-ci`.

Worth mentioning are two very useful tools that aid in writing *Journeys* and locating controls:

- The UI5 Journey Recorder[38] is capable of – just as the name suggests – recording the navigation through a UI5 app. It can output coding suitable for OPA5 and wdi5 (see the next chapter) of the recorded journey.

- The UI5 Test Recorder[39] allows for interactively finding a Control in a UI5 app and getting the best possible locator for it. Note that when using the UI5 Tooling for serving your app, you need to include the npm module `@openui5/sap.ui.testrecorder` and some dependencies in the `ui5.yaml` (Listing 2-43)!

---

[38] https://github.com/ui5-community/ui5-journey-recorder
[39] https://ui5.sap.com/sdk/#/topic/2535ef9272064cb6bd6b44e540
2d531d.html

Listing 2-43. Sample `ui5.yaml` excerpt for the UI5 Test Recorder

```
framework:
  name: OpenUI5
  version: 1.112.0
  libraries:
    - name: sap.ui.core
    - name: sap.m
    - name: sap.f
    - name: sap.ui.layout
    - name: themelib_sap_horizon
    - name: sap.ui.testrecorder
    # all of the below are dependencies for the Test
    Recorder...
    - name: themelib_sap_fiori_3
    - name: sap.ui.table
    - name: sap.tnt
```

End-to-End Tests

For simulating how a user operates your UI5 application, probably including navigating to it from a non-UI5 app, End-to-End tests come into play. For UI5, the open source tool wdi5[40] is used for that. It is an extension to WebdriverIO and capable of remote-controlling a wide range of browsers on Windows, macOS, and Linux.

Setting up wdi5 can comfortably be done in a UI5 app via `npm init wdi5@latest`. This will bootstrap a minimal setup with Google Chrome as the remote-controlled browser target. After the `npm init`, tell wdi5 via its config file in /apps/openui5/*wdio.conf.js* to look for test files ending in `test.js` in the proper directory:

`specs: ["./uimodule/webapp/test/e2e/**/*.test.js"]`

---

[40] https://ui5-community.github.io/wdi5

wdi5 test files can be formulated in either Mocha,[41] Cucumber,[42] or Jasmine.[43] All dialects specify a syntax to make the test file human-readable. For this chapter, we'll use Mocha. A minimal wdi5 test file written in Mocha looks like Listing 2-44.

Listing 2-44. Mocha-flavored wdi5 test file example

```
describe("read scenario", () => {
  it("should find the 'Blue Moon' as the first beer",
  async () => {
    const firstListItem = await browser.
    asControl(listSelector).getItems(0);
    const firstListItemTitle = await firstListItem.getTitle();
    expect(firstListItemTitle).toEqual("Blue Moon Belgian
    White");
  });
});
```

describe denotes a suite consisting of a number of tests, the its. The central API to locate a UI5 control is browser.asControl($locator). Given the asynchronous nature of browser interactions, every operation in wdi5 needs to be awaited.

Looking closer at Listing 2-44, you'll notice the familiar getItems method on a UI5 List Control and the getTitle of the Standard List Item. wdi5 enables the usage of a retrieved UI5 Control's API at design time, even though the test runs in the Node.js context (and the UI5 app – obviously – in the browser context). This API alignment reduces the friction to use wdi5 when coming from OPA5.

---

[41] https://mochajs.org/
[42] https://cucumber.io/docs/guides/
[43] https://jasmine.github.io/

Additionally, wdi5's selectors are intentionally kept **compatible with those of OPA5** to allow for seamless transition between tools for the developer. The `listSelector` in Listing 2-45 looks identical in wdi5 and OPA5, with the wdi5-specific wrap in a `selector` property.

Listing 2-45. Identical locators in OPA5 and wdi5

```
selector: {
  id: "container-com.apress.openui5---BeerList--beerList"
}
```

Check out the many selector examples in our repo at /apps/openui5/*uimodule/webapp/test/e2e/_selectors.js* that are usable by both OPA5 and wdi5.

Clearly, the line between OPA5 and wdi5 is blurred. By intention, that enables UI5 developers to get familiar with both via shared learnings. By personal preference, this gives the developer the flexibility to choose the tool of their liking.

While OPA5 runtimes are faster, wdi5 allows for parallel execution of tests. Screen captures and file uploads and downloads are only possible with wdi5, yet OPA5 operates in the same browser scope that the app does and doesn't require the overhead of an API alignment – albeit there are test capabilities uniquely enabled by wdi5: aforementioned screenshots and file handling, authentication for an app, and testing across UI5 and non-UI5 apps, just to name a few.

Now that you've cloned our repo and installed all dependencies, here's how to run the End-to-End tests:

1. Open a Terminal and change into the app folder with `cd apps/openui5`. Start the UI5 Tooling's web server with `npm start`.

2. Open another Terminal and again change into the
app folder with cd apps/openui5. Then execute
the wdi5 tests in an automated fashion just like a CI
would via npm run wdi5.

You'll see two scenarios running with Google Chrome: a "read
scenario" validating the list of beers and a "write scenario" adding a new
beer to the list.

By the way, the same two tools mentioned in the "Integration Tests"
section that help with finding UI5 controls programmatically (UI5 Test
Recorder) and recording the usage of a UI5 app (UI5 Journey Recorder)
also work for wdi5. Their code generation option includes an export to
wdi5 as well!

Deploy Continuously

While it is OK in the beginning of a development phase to deploy a UI5
app manually, sooner rather than later the requirement for automated
deployments comes up – it is less error prone and not dependent on
specific people. The process of automated deployments triggered by a
certain event is commonly referred to as Continuous Deployment (CD).

The first major prerequisite for CD is Continuous Integration (CI) –
running automated tests based on a certain trigger. The setup for the
different test scopes was handled previously in the chapter. Only when the
required number of tests passes should a subsequent deployment happen
automatically.

The second major prerequisite for CD is a build phase. In it, the
UI5 application sources are transpiled, optimized, and bundled for the
deployment target. The build phase might even be a requirement for the
CI phase in order to really test on the sources that will potentially go into
production rather than on the development artifacts.

Both the triggers and the rules for CI and CD vary greatly from project to project. One might run CI on every newly pushed branch into the version control system, while deployment happens only on changes to the main branch. Another might run CI only on successfully reviewed pull/merge requests, but deploy every pull/merge request automatically for easy availability of the changed code. So we won't cover the various combinations of CI/CD possible, but rather recommend an approach for establishing CD at all: cut processes locally, then tailor them to the central DevOps environment.

Cut Locally, Move to Central

No matter what DevOps platform is in use (GitHub, GitLab, Bitbucket, Jenkins, etc.), a reliable and proven strategy is to first get the build and deployment process running locally. This doesn't entail the entire process, but central parts of it. For example, the automated cloning/checkout of the repo doesn't need local triage. Instead, the DevOps platform should be trusted with "bread and butter" operations.

What should first be tested locally is the build phase: Are you using the same Node.js version that is running in the deployment environment? Do the packaged sources match the expectations and requirements of the deployment target? Is all tooling required for the build specified as a dependency in the repository? Or does the DevOps platform only allow tool usage via containers (Docker)? So the build might need a specific image available from a specific container repository that needs to be put together prior to any possible build.

If the build phase runs fine locally, move on to test the deployment phase locally next. Again, is all tooling required for the deployment available as a dependency in the repository? Or does it require certain Docker images? Can the credentials necessary for the deployment be put into DevOps platform's secrets or do they have to be generated on the fly during the publication process?

Also recommended here is to think of proper communication channels for both failed and successful deployments. Is an instant messaging or chat channel available for use? Is email set up properly on the DevOps platform? Do text messages need to be sent via cellular networks? Does a failed deployment need to be automatically retriggered a certain amount of times?

All of these can and should be tested locally first. If the deployment with its major steps work fine on your machine, then start porting both build and deployment processes over to the DevOps platform. Take it one step at a time: start with the build first, then continue with the deployment steps. And don't underestimate the time necessary for this: even when things work perfectly locally, the move to the central platform can take more time than expected! Make enough time for this early in your project – but start getting Continuous Integration and Deployment going as soon as possible!

The @sap/approuter As the Central Module

A UI5 application is "just another web application." So generally speaking, it can be deployed to whatever environment that offers web server capabilities.

Typical deployment targets in the SAP world are on-premise SAP systems, the SAP Build Work Zone, standard edition[44] (formerly known as SAP BTP Launchpad service) with the HTML5 application repository,[45] the SAP BTP Kyma,[46] or Cloud Foundry runtime[47] in general. In many

---

[44] www.sap.com/products/technology-platform/workzone.html

[45] https://help.sap.com/docs/btp/sap-business-technology-platform/html5-application-repository

[46] https://discovery-center.cloud.sap/serviceCatalog/kyma-runtime?region=all

[47] https://discovery-center.cloud.sap/serviceCatalog/cloud-foundry-runtime/?region=all

scenarios, the @sap/approuter,[48] or *approuter* in short, plays a central role, albeit to various extents. That's why for our example app, we'll focus on deploying a stand-alone UI5 application together with an approuter to a BTP Cloud Foundry space.

In our scenario, look at the approuter as the extended security detail for the app. It not only acts as the web server delivering the app to clients but also applies the authentication and authorization configured in BTP to the app, plus destination handling. The approuter configuration in *xs-app. json* specifies who gets access to your app (authentication) and what the person is allowed to see/do in your app (authorization).

Here's what's to add to */apps/openui5* for bringing the approuter into the game with a minimal configuration (Listing 2-46).

Listing 2-46. The deployment artifacts

```
├── approuter
│   ├── package.json
│   └── xs-app.json
└── mta.yaml
```

In the dedicated approuter directory, package.json only holds the necessary package dependency to @sap/approuter itself. And a start script for the approuter with default options (Listing 2-47).

Listing 2-47. Approuter initiation

```
{
  "name": "com.apress.openui5",
  "dependencies": {
    "@sap/approuter": "latest"
  },
```

---

[48] www.npmjs.com/package/@sap/approuter

```
"scripts": {
  "start": "node node_modules/@sap/approuter/approuter.js"
}
}
```

xs-app.json describes how the approuter should protect the app and what HTTP traffic to distribute (routes) where (Listing 2-48).

Listing 2-48. Approuter configuration

```
{
  "welcomeFile": "index.html",
  "authenticationMethod": "route",
  "routes": [
    {
      "source": "^(.*)$",
      "target": "$1",
      "authenticationType": "xsuaa",
      "localDir": "webapp/"
    }
  ]
}
```

The only route maps incoming traffic to / onto webapp/ so the domain the approuter serves can omit webapp in its path. If deployed at https:// rusty-richie-unix.cfapps.eu20.hana.ondemand.com, the domain serves the webapp folder as its root with the help of the approuter. So no need to call https://rusty-richie-unix.cfapps.eu20.hana.ondemand.com/ webapp/ to access our app.

authenticationType is set to xsuaa, so whatever is configured in terms of Identity Provider and Authentication protocol in the BTP will be picked up by the app.

mta.yaml specifies the infrastructure configuration for the app on BTP Cloud Foundry (Listing 2-49).

Listing 2-49. Cloud Foundry deployment descriptor `mta.yaml`

```
ID: ui5-approuter
_schema-version: 3.2.0
version: 1.0.0
parameters:
  enable-parallel-deployments: true

modules:
  - name: "com.apress.openui5"
    type: approuter.nodejs
    path: ./approuter
    parameters:
      disk-quota: 256M
      memory: 256M
    requires:
      - name: html5-uaa
resources:
  - name: html5-uaa
    type: org.cloudfoundry.managed-service
    parameters:
      service: xsuaa
      service-plan: application
      config:
        xsappname: "ui5-approuter-uaa"
        tenant-mode: dedicated
```

A single module `com.apress.openui5` is deployed as type `approuter` that requires an instance of the "User Authentication and Authorization" (uaa) service on BTP for, well, authentication and authorization.

With all the deploy- and build-time artifacts in place, it is time to cut the actual build. Install the "Cloud MTA Build Tool" `mbt` as a development dependency in */apps/openui5*: `npm i --save-dev mbt`.

Then we specify an npm script to

1. Optimize the UI5 app for deployment:

```
ui5 build self-contained --all --config=
uimodule/ui5-deploy.yaml --clean-
dest --dest dist
```

   This does a self-contained build of the webapp folder, configured via ui5-deploy.yaml, into a folder dist and first cleans any existing files from dist.

2. Move the self-contained build artifacts from ./dist into ./approuter/webapp and then use the mbt for building the final deployment artifact:

```
mv dist approuter/webapp && mbt build
```

Listing 2-50 shows the excerpt from package.json bringing it all together.

Listing 2-50. npm build script for deployment

```
"scripts": {
    "build:deploy": "run-s build:uimodule:deploy build:mtar",
    "build:uimodule:deploy": "ui5 build self-contained
    --all=true --config=uimodule/ui5-deploy.yaml --clean-dest
    --dest dist",
    "build:mtar": "mv dist approuter/webapp && mbt build"
}
```

Running npm run build:deploy generates the so-called multitarget application (mtar) file at *mta_archives/ui5-approuter_1.0.0.mtar* that is ready to be pushed to Cloud Foundry!

Install the Cloud Foundry CLI[49] if you haven't yet, along with the multiapps plugin[50] for handling mtar files.

Log in to your Cloud Foundry account via cf login.

Select the appropriate space to deploy the UI5 app to.

And then... (Listing 2-51).

Listing 2-51. Log of a successful cf deploy

```
$> cf deploy mta_archives/ui5-approuter_1.0.0.mtar -f
Deploying multi-target app archive mta_archives/ui5-
approuter_1.0.0.mtar in org xxx / space yyy as you@
example.org...

Uploading 1 files...
  /Users/you/apps/com.apress.openui5/mta_archives/ui5-
  approuter_1.0.0.mtar
OK
Operation ID: 25bdd2b1-d70c-11ed-a3c2-eeee0a9d6eeb
Deploying in org "xxx" and space "yyy"
Detected MTA schema version: "3"
Detected deployed MTA with ID "ui5-approuter" and
version "1.0.0"
#...
Creating application "com.apress.openui5" from MTA module "com.
apress.openui5"...
# ...
Starting application "com.apress.openui5"...
Application "com.apress.openui5" started and available at "xxx-
yyy-com-apress-openui5.cfapps.eu20.hana.ondemand.com"
```

Point your browser to the logged URL – up, up, and away!

---

[49] https://docs.cloudfoundry.org/cf-cli/install-go-cli.html
[50] https://github.com/cloudfoundry/multiapps-cli-plugin

The Road from Here

The previous tutorial provided a rough introduction and taught some of the most frequently used concepts of UI5. However, this is just a starting point, and we recommend you continue building other web apps on your own. There are also tutorials in the official documentation that will regularly be updated when new versions are released:

- Get Started: Setup, Tutorials, and Demo Apps[51]

- SAPUI5 Walkthrough Tutorial[52]

- SAPUI5 Tutorial Missions in the Tutorial Navigator[53]

Besides these tutorials, we recommend starting your project without an existing sample solution. Popular ideas for these kinds of projects are

- A to-do list application

- A program that converts temperatures, or other units, between Fahrenheit and Celsius

- A basic calculator

- A web app that suggests random passwords

- A program that retrieves and displays data from an API

When you start with one of these, you'll soon reach the point when you need to find more information about concepts, browse all available controls, or get to know the API of controls. The SAPUI5/OpenUI5 Demo Kits will provide the needed tools for you, such as the SAPUI5

---

[51] https://ui5.sap.com/#/topic/8b49fc198bf04b2d9800fc37fecbb218
[52] https://ui5.sap.com/#/topic/3da5f4be63264db99f2e5b04c5e853db
[53] https://developers.sap.com/tutorial-navigator.
html?tag=programming-tool%3Asapui5&tag=tutorial%3Atype%2Fmission

Documentation[54] (including Best Practices for App Developers[55]), SAPUI5 API Reference,[56] and the SAPUI5 Controls Library.[57] The SAPUI5 Live Editor[58] is a (Read-Evaluate-Print-Loop) REPL-like, in-browser sandbox that you can use for testing, and the SAPUI5 Icon Explorer[59] provides an overview of all available icons that come with the framework. And finally, you can have a look at the UI Integration Cards[60] when you need controls to visualize analytical information.

Analogous, you can get to the respective OpenUI5 resources when you replace the `https://ui5.sap.com/*` domain with `https://sdk.openui5.org/*`.

The great innovation in this field is not only coming from SAP directly. The active community around this technology is also pushing in the right direction. These community projects include applications, libraries, custom controls, tooling extensions, middleware, tasks, or easy-ui5 generators. And Best of UI5[61] is the central place where you can get to know them.

UI5 for Enterprise Applications

Overall, UI5 addresses all requirements discussed in the previous chapter. It provides a consistent user experience and a palette of enterprise UI elements that strictly follow the SAP Fiori design system. This is not

---

[54] https://ui5.sap.com/#/topic
[55] https://ui5.sap.com/#/topic/28fcd55b04654977b63dacbee0552712
[56] https://ui5.sap.com/#/api
[57] https://ui5.sap.com/#/controls
[58] https://ui5.sap.com/#/liveEditor
[59] https://ui5.sap.com/test-resources/sap/m/demokit/iconExplorer/webapp/index.html
[60] https://ui5.sap.com/test-resources/sap/ui/integration/demokit/cardExplorer/index.html
[61] https://bestofui5.org/

surprising as SAPUI5 predates the SAP Fiori standard. The framework was built with security concerns kept on top of mind and makes it for app developers as easy as possible to build secure web applications that are resistant to the most common attack vectors. UI5 is compatible with all modern browsers and only drops support for browsers if that has a strategic reason, as was the case with the dropped support for Internet Explorer 11[62] in early 2021.

SAPUI5 grew much in its 15-year-old history and brought a number of innovations to its user base. They upgraded its underlying rendering engine to be more performant and secure, offered multiple options for testing and TypeScript support, and made the core leaner by moving out dependencies like jQuery, while other things remained constant to avoid unnecessary disruption and ease the maintainability of built applications. This includes the core principles around XML or JSON views, models, the data binding connecting them, controllers, controls, i18n support, and accessibility. Individual components such as controls and controllers can be extended if business needs require this. Coming from the same vendor, SAP, the framework makes it easy to integrate with SAP backend systems in the cloud or on-premise. This integration covers data exchange and the platform on which the applications are hosted. It's important to note UI5 apps can also run on any web server and connect to any data source.

SAPUI5, and its open source sibling, OpenUI5, balanced a trade-off that gives developers the freedom to build any kind of application with great support for enterprise applications. Like every trade-off, this one could not and is not without downsides. Some software projects require more freedom, flexibility, and maybe better performance. So the developers might think they are better off with other rendering frameworks as they only desire to build applications that appear SAP Fiori compliant. This additional freedom usually results in a larger investment in the projects but can make sense for multiple reasons. At the same time,

---

[62] https://twitter.com/openui5/status/1349612212168646657

there are developers who would like to delegate even more tasks to their frontend framework. They want to invest as little time and resources as possible in writing new applications. There could be multiple reasons for this, either they just want to prototype a new process, need to cover a business scenario that only occurs rarely and is not worth spending a lot of time on, or they rapidly need to adapt to a new business situation and rather need a quick solution than a perfect one. They think that SAP Fiori already provides the floorplans they want, and now they only want to connect them to a data source to work out of the box. The good news is that SAP also offers options for these developers. And we'll cover them in the next chapter.

CHAPTER 3

Colossal: SAP Fiori Elements

In this chapter, we will unveil the immense power of SAP Fiori elements and explore the fundamental concepts behind the OData protocol and its role in enabling seamless integration between frontend and backend services.

We'll then dive deep into SAP Fiori elements, a technology that simplifies UI development. We will highlight its various use cases and benefits and walk you through a hands-on tutorial, guiding you step by step through bootstrapping a project with SAP Fiori elements.

For developers already familiar with UI5 freestyle apps, we'll demonstrate how to incorporate SAP Fiori elements selectively into existing applications.

To wrap things up, we'll explore the road ahead and the future possibilities of using SAP Fiori elements in enterprise applications.

With the stage set, let's get started.

What Is SAP Fiori Elements?

SAP Fiori elements is a development toolkit that allows developers to build SAP Fiori apps more quickly and efficiently. It provides a set of predefined UI components and templates that can be used to create common business floorplans and related functionality. It leverages annotation-based configuration, similar to Java EE, to keep the implementation work at a minimum and, therefore, to stay as close to the SAP Fiori design system as possible. By using SAP Fiori elements, developers can focus on building the business logic of their apps instead of spending time on low-level UI implementation details.

Because of its narrow focus on enterprise applications and data retrieval from SAP backends, this toolkit is not open source and not available in OpenUI5. Having said that, there is an SAP Fiori elements add-on for OpenUI5[1] that includes selected floorplans to make SAP Fiori elements available to a wider community. While it's possible to use this add-on in OpenUI5 and SAP considers it freeware,[2] it doesn't mean the @sap/open.fe[3] add-on itself is open source or published under an open source license.

SAP Fiori elements themselves are implemented with SAPUI5 and, therefore, also include features such as data binding via models. It was formerly also known as "Smart Templates" and started to get popular in 2017 because it allowed developers to easily connect their UI to ABAP-based backend data sources and customize the behavior and appearance of their apps. Today, SAP Fiori elements is a one-stop shop for all apps that want to connect to any SAP-related data source to display an overview page, list report, object page, or any other floorplan introduced in Chapter 1. These pages come already fully implemented by the toolkit, and the developers only need to hook them to a data source. There's no need to go to the whiteboard to come up with a design, a keyboard navigation pattern, draft or message handling, or localization. Therefore, SAP Fiori elements cannot only be used to minimize the implementation cost but

---

[1] https://blogs.sap.com/2020/12/21/
now-available-sap-fiori-elements-add-on-for-openui5/
[2] https://tools.hana.ondemand.com/sap-freeware-license.txt
[3] www.npmjs.com/package/@sap/open.fe

also the maintenance cost while managing the risk of regressions. The only thing that needs to be done by the developer is choosing the annotation. Popular configurations are, for example, whether the line items of a report are selectable or not or whether items can be edited, removed, or just viewed. To avoid feature limitations, developers are free to go beyond the configuration options of the floorplans and embed custom UI5 fragments in the floorplan.

You might already guess that the data the SAP Fiori elements app (Figure 3-1) visualizes needs to come from a standardized source.

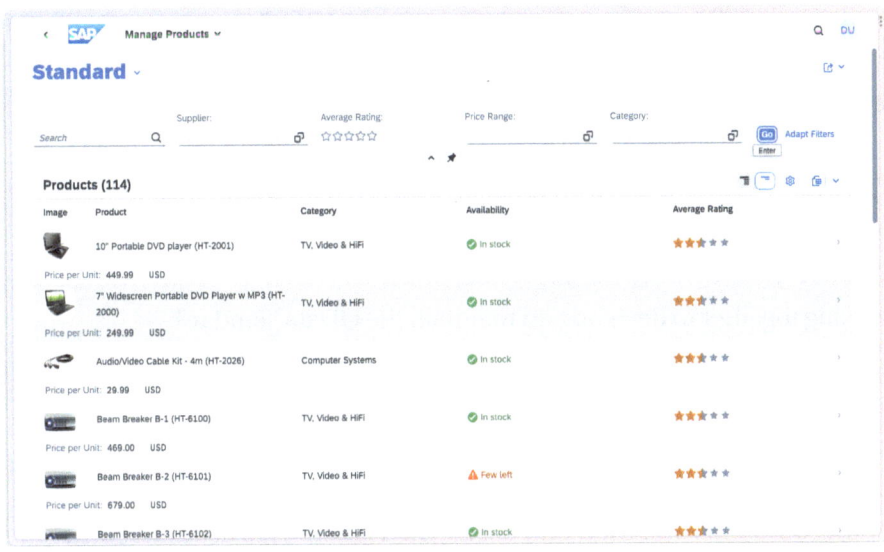

Figure 3-1. *A sample list report application showing various products*

This standardization is required so that the generic user interfaces know exactly how data objects are structured and linked to other objects, for example, for navigation purposes, to sort and filter and manipulate the data. Think of a list of products with multiple properties such as product name, category, and stock information on the landing page. When you click a product, you'll see a detailed object page describing the products with recent customer reviews, stock forecast, generated revenue in the past month, and all pending orders. It's not trivial to organize such information so that the user interface can present it correctly. And the Open Data (OData) protocol, which is already supported by models in UI5, is the solution to this problem.

Introducing the Open Data Protocol

The Open Data (OData) protocol is a standard for creating and consuming data APIs. OData is a RESTful API protocol, meaning that it is designed to be used over HTTP and is based on the principles of REST (representational state transfer). However, OData adds functionality on top of REST and provides several additional features that can make it easier for developers to create and consume data via APIs. The open standard is maintained by the OASIS (Organization for the Advancement of Structured Information Standards) OData Technical Committee, which comprises a group of individuals from various organizations, such as Microsoft, SAP, and Dell. These individuals are working together to develop and maintain the OData standard.

The protocol is used by a wide range of systems and platforms for creating and consuming data APIs. It's commonly used to expose data and functionality from enterprise systems, such as ERP and CRM systems, to external applications and services. OData provides many benefits for these systems and their end users when it comes to accessing and manipulating data. The protocol is extendable and supports filtering, sorting, navigation, and pagination operations. This can be particularly useful as enterprise software systems often deal with large volumes of data. It also supports a number of different data formats, including JSON and XML, which can make it easier to integrate with a wide range of systems and technologies. This can be beneficial as it allows the backend to expose its data from various sources, including databases, file systems, and cloud-based services, in a way that is compatible with a wide range of client applications. There are multiple versions of the protocol, and the functionality and syntax may differ between versions. It's essential for developers to be aware of the version they are working with and its specifics. While OData version 2 is still in use for some SAP products, we'll focus on the newer version 4[4] for the remainder of this book.

From a technical point of view, OData defines a data model for representing the entities and relationships of an API. It describes how data can be represented as entities, with properties and navigation properties, and how those entities can be related to each other. It supports a set of operations such as Create Read Update Delete (CRUD), and it's extensible to support more custom operations; these operations are mapped to standard HTTP methods (GET, POST, PUT, DELETE). For data access, it provides a set of query options to filter, sort, and paginate data, such as $select, $filter, $orderBy, and $top.

You could also say that OData simplifies the life of API developers because they don't need to implement the CRUD and query operations for each entity and their relations any longer. Instead, they can focus on the definition of their data model and rely on the OData protocol and helper libraries that generate the API for client access. So in some way, OData does for API developers what SAP Fiori elements does for the frontend developers, which makes them a natural fit.

Let's think of the data model of beers and breweries as shown in Figure 3-2.

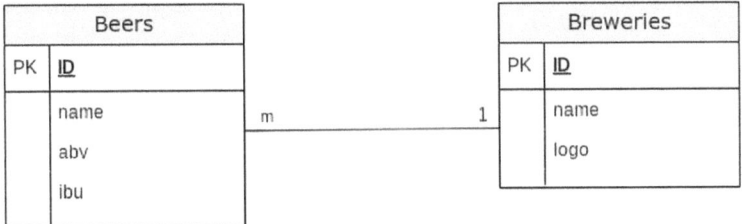

Figure 3-2. *The data model we'll use in this book*

---

[4] https://docs.oasis-open.org/odata/odata/v4.0/odata-v4.0-part1-protocol.html

The query in Figure 3-3 retrieves a list of beers from an endpoint called beershop and filters the results to only include the name and abv fields, ordered by abv in ascending order, and only including beers with an abv value greater than 3.

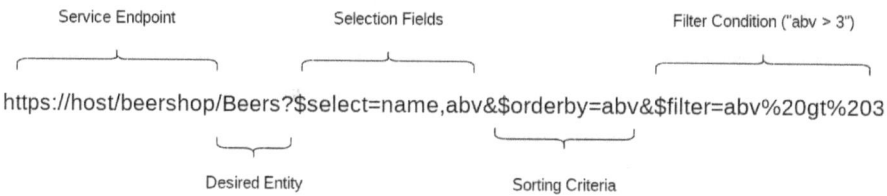

Figure 3-3. *Structure of an OData request*

Northwind is a popular sample OData service that exposes sample data and is often used as a test service for developers working with OData. Unfortunately, the sample service for OData version 4 isn't strictly following the protocol and returns, according to SAP, invalid responses.[5] So far, Microsoft, which maintains the sample service, has not replied.[6] Hence, we do not recommend using the sample service for the development of SAP Fiori elements apps. For simple testing, we will use the "beershop" data model that we already saw earlier. Other OData services you might want to consider are the bookstore[7] and sflight[8] sample services that have been generated with the SAP Cloud Application Programming Model.[9]

---

[5] https://github.com/SAP/openui5/issues/2014

[6] https://github.com/OData/ODataSamples/issues/216

[7] https://github.com/SAP-samples/cloud-cap-samples

[8] https://github.com/SAP-samples/cap-sflight

[9] https://cap.cloud.sap/docs/

SAP Fiori Elements Configuration

Now that you know what SAP Fiori elements are and *that* their data comes from an OData service, you probably wonder *how* it does that. There are actually two *how questions* you could ask:

- How does the SAP Fiori elements app know which floorplans to use for which page?

- How does the SAP Fiori elements app know which API data to use for which UI element of a page?

We previously learned each SAP Fiori elements app is implemented with UI5, and, therefore, it builds on the same concepts and has an application descriptor file – the manifest.json. This file declares all important properties of an application, among others, also the routing configuration. The app uses this routing configuration to determine which UI to use for a particular page. All the page specifications, or targets as they are called here, are also defined in this file. In the previous section, we used the object in Listing 3-1 to define a target for the router.

Listing 3-1. A sample routing entry

```
{
  "TargetBeerAdd": {
    "type": "View",
    "viewType": "XML",
    "viewId": "BeerAdd",
    "viewName": "BeerAdd"  }
}
```

We didn't use the property type in the previous chapter because it wasn't necessary. UI5 knew, based on the other properties (viewId and viewName, for example), it was a view and could pick the right file to render the content. Using the same mechanism, we can define a target that isn't

131

implemented as a view but as a reusable SAP Fiori floorplan template. We call this a Component type that is specified with its name that belongs to the library sap.fe.templates. Listing 3-2 shows how component-specific configuration can be added manually to this file. Later, in the hands-on section, we'll see how tools can help us to manage this configuration.

Listing 3-2. Component-type routing target

```
{
  "OrdersList": {
    "type": "Component",
    "id": "OrdersList",
    "name": "sap.fe.templates.ListReport",
    "options": {
      "settings": {
        "entitySet": "Orders",
        "variantManagement": "Page",
        "navigation": {
        "Orders": {
          "detail": {
            "route": "OrdersObjectPage"
          }
        }
      }
    }
  }
}
```

Web applications leveraging the SAP Fiori elements toolkit can have their own controllers and views, but they are not required to do so. Therefore, these directories are not mandatory for such applications. With this, the first question is answered: **the SAP Fiori elements app knows about the used floorplans and their navigation paths via the configuration in the manifest.json.**

The answer to the second question is also slightly related to the manifest file. The SAP Fiori elements app uses the OData v4 model to connect to the service and retrieve data. The model is typically configured in the manifest.json file and is specified by the endpoint URL of the API and any authentication details required to access the API. Once the OData model is configured, the app can use it to retrieve data from the API and bind it to the UI elements of a page. The big difference to normal UI5 apps, using the data binding syntax, is that we don't write the views, and, therefore, we cannot use the binding syntax we already know. SAP Fiori elements, on the other hand, use **OData annotations** to provide additional information about the data retrieved. The annotations can either be part of a single web app, come from the backend for multiple web apps at the same time, or be a mix of these two sources. Annotations are typically defined in an entity data model XML (EDMX) file. The file contains information about the entities, properties, associations, and other elements that make up the data model. Listing 3-3 shows an example of an EDMX file that is used by the OData protocol.

Listing 3-3. The structure of an OData EDMX file

```
<edmx:Edmx xmlns:edmx="http://docs.oasis-open.org/odata/ns/
edmx" Version="4.0">
    <edmx:Reference Uri="https://sap.github.io/odata-
    vocabularies/vocabularies/UI.xml">
        <edmx:Include Namespace="com.sap.vocabularies.UI.v1"
        Alias="UI"/>
    </edmx:Reference>
    <edmx:Reference Uri="/serviceurl/$metadata">
        <edmx:Include Namespace="BeershopService"/>
    </edmx:Reference>
    <edmx:DataServices>
```

```
<Schema xmlns="http://docs.oasis-open.org/odata/ns/edm"
Namespace="local">
    <Annotations Target="BeershopService.Beers">
        <!-- ... -->
    </Annotations>
</Schema>
</edmx:DataServices>
</edmx:Edmx>
```

The top-level element is the edmx:Edmx element, which defines the version of the OData protocol and the namespaces that are used in the file.

The edmx:Reference element is used to reference other vocabularies or metadata files that are used in the EDMX file. In this example, the first edmx:Reference element references the UI vocabulary and includes it with the alias UI, and the second edmx:Reference element references the metadata of the service and includes it with the namespace BeershopService.

The edmx:DataServices element is used to define the data services that are provided by the OData service. Inside this element, there is a <Schema> element that defines the schema and default namespace of the data service.

The <Annotations> element is used to define annotations for the entities of the service. In this example, the annotations are targeted at the BeershopService.Beers entity. The content of the annotations is not provided in the example, but we'll add some later.

Annotations can not only be used to bind UI elements to the data but also to customize the behavior and appearance of UI elements. Additionally, they can also provide additional information such as

- Date and currency formatting

- Navigation properties between entities

- Grouping of various fields

- Value help

- Semantic information about the data, like data fields used as keys, or identifiers, that could be used for filtering, sorting, or grouping

- And many other things

The UI.DataField annotation is used to indicate that a particular property represents a data field in the OData service. It can be used to map a property to a specific column in a database table or to a field in an XML file.

UI.HeaderInfo is an annotation that is used to provide additional information about the entity or property that is used to generate the user interface. It can be used to specify the label for the field, the type of the field, and other information that is used to generate the user interface.

The UI.LineItem annotation is used to indicate that a particular property is a line item in a collection of entities. It is often used with the UI.DataField and UI.HeaderInfo annotations.

Let's see how texts and labels for fields are represented in these annotations (Listing 3-4).

Listing 3-4. OData XML annotation

```
<Annotations Target="BeershopService.Beers">
  <Annotation Term="UI.LineItem">
    <Collection>
      <Record Type="UI.DataField">
        <PropertyValue Property="Label" String="name"/>
        <PropertyValue Property="Value" Path="name"/>
      </Record>
    </Collection>
  </Annotation>
```

```
<Annotation Term="UI.HeaderInfo">
  <Record>
    <PropertyValue Property="Name">
      <Record Type="UI.DataField">
        <PropertyValue Property="Value" Path="name"/>
      </Record>
    </PropertyValue>
    <PropertyValue Property="ABV">
      <Record Type="UI.DataField">
        <PropertyValue Property="Value" Path="abv"/>
      </Record>
    </PropertyValue>
  </Record>
</Annotation>
</Annotations>
```

These are Fiori-specific annotations defined in the SAP Vocabularies.[10] With them, you can customize the behavior and appearance of the SAP Fiori elements web app. But don't worry, you don't need to learn them all by heart. There are a number of tools that can help you during development, and we'll get to know them in the next section.

Learning by Doing

Same as in the previous section, we will solidify the mentioned concepts by building a small demo application. We already mentioned that one of the bigger benefits of SAP Fiori elements development is that you do not need to write a lot of code, and therefore little maintenance work is needed. This lack of code also makes it a bit hard to cover how to build applications with it. Hence, this section mostly highlights which tools help with which tasks.

---

[10] https://github.com/SAP/odata-vocabularies

Before you start building, please ensure you set up your development workspace[11] as instructed in the previous chapter. The demo application will display a number of beers that are defined in the beer shop sample application.[12] The data model has been predefined, and now your boss asks you to build a prototype that can display all beers in a list. This list shall be filterable to immediately see all names and breweries of the beers. Furthermore, this list should also navigate to a detailed view that shows not only the ABV and IBU properties but also the logo of the brewery. And ideally, this web app should be implemented today and reuse common UX standards that can be understood without training. A responsive design would also be nice so that the site adjusts to the screen size of the device being used, whether it's a desktop or mobile device. Unfortunately, your team is understaffed, as always, and you only have some time after the lunch break. The good news is that SAP Fiori elements can save the day.

Bootstrap the Project

Our demo application shall be based on the beer shop data model we talked about previously. **Run the following commands** to download the edmx file that defines this service, so we can focus on the web application project itself afterward.

All the terminal commands used in this section are for Unix systems like macOS and Linux. If you're using Windows, you'll need to use the Windows Subsystem for Linux to access those same commands. Windows Subsystem for Linux provides a Linux-compatible environment that you can use to run many Unix tools and applications on your Windows machine.

---

[11] Chapter 2, "Setting Up Your Workspace".

[12] github.com/Apress/SAP-UI-Frameworks-for-Enterprise-Developers-by-Marius-Obert-Volker-Buzek

```
mkdir fiori-elements-demo
cd fiori-elements-demo
curl https://raw.githubusercontent.com/Apress/SAP-UI-
Frameworks-for-Enterprise-Developers-by-Marius-Obert-Volker-
Buzek/main/apps/fiori-elements/webapp/localService/metadata.xml
> metadata.xml
code . # this opens VS Code from the command line
```

In VS Code, open the Command Palette and find the command "Fiori: Open Application Generator" (Figure 3-4). This wizard came with the VS Code extension "SAP Fiori Tools Extension Pack" you installed when setting up the development environment.

Kindly note that the majority of the code presented in this section has been automatically generated using version 1.9.6 of the SAP Fiori Tools Extension Pack. It is important to be aware that potential discrepancies may arise in both the generated code and the resulting user interface when utilizing future versions of the extension pack.

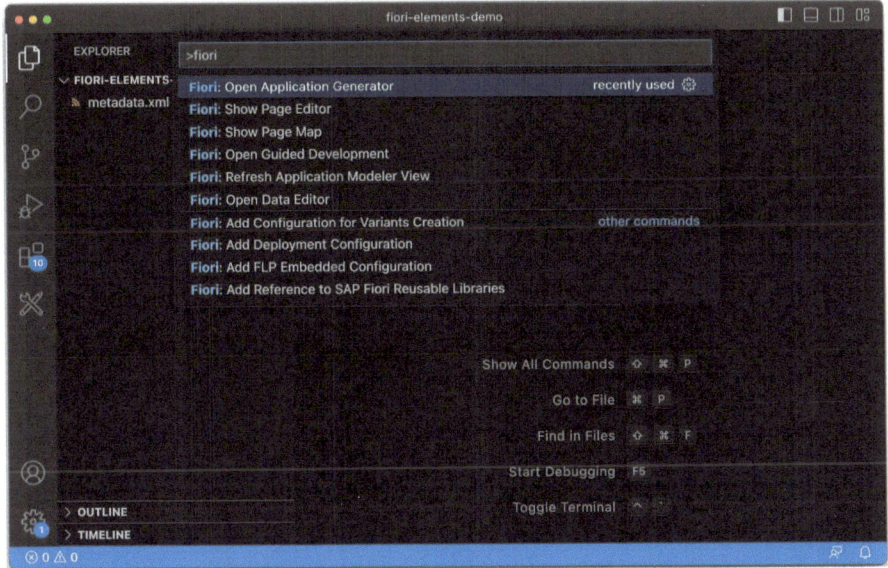

Figure 3-4. *VS Code Command Palette showing "Fiori: Open Application Generator"*

A template wizard will open and ask you for the application type; select SAP Fiori here and the template. As we want to display a list of entities, we should choose the List Report Page before we continue with **Next** (Figure 3-5).

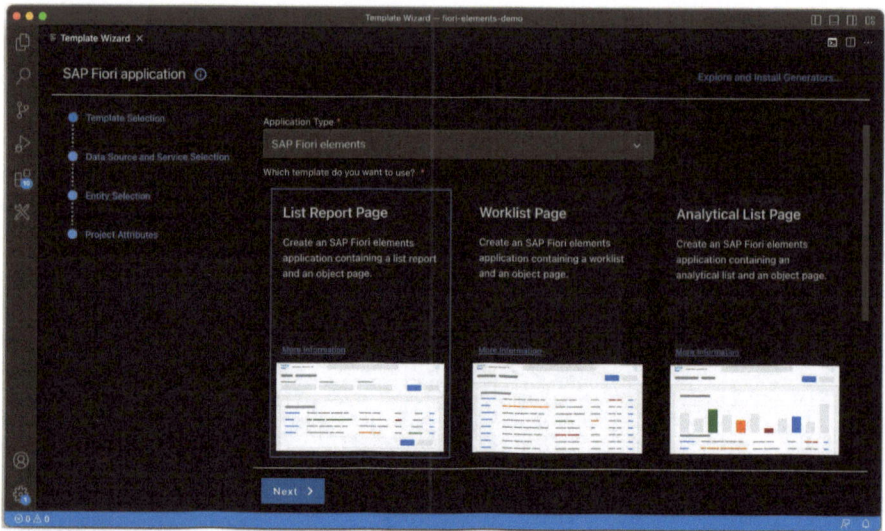

Figure 3-5. *The template wizard to bootstrap a new project*

When asked for a data source, select "Upload a Metadata Document" referring to the EDMX file we just downloaded with the command line in the new folder (Figure 3-6).

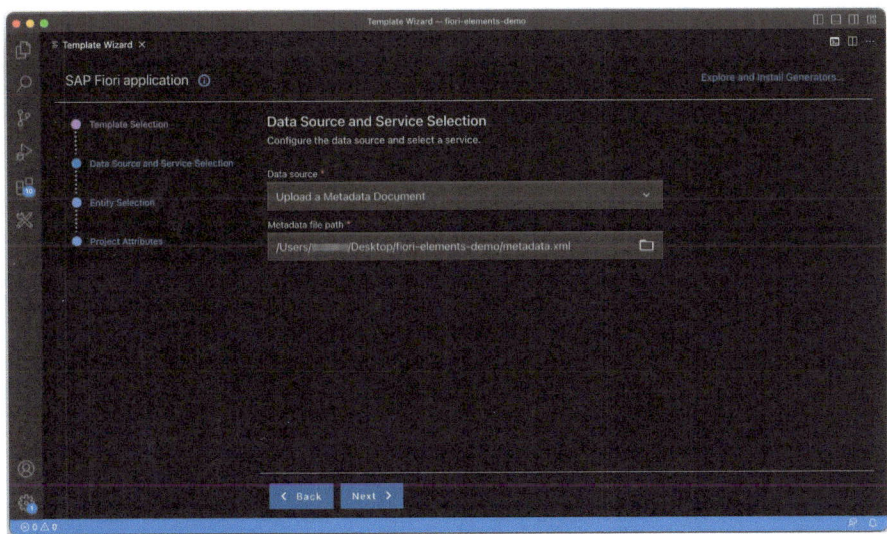

Figure 3-6. *Selection of the data model*

The "Main entity" we want to list is the "Beers" entity. Keep the other default values on this wizard page and click **Next**.

In the last wizard step, define the project attributes as shown in Figure 3-7 before you **Finish**.

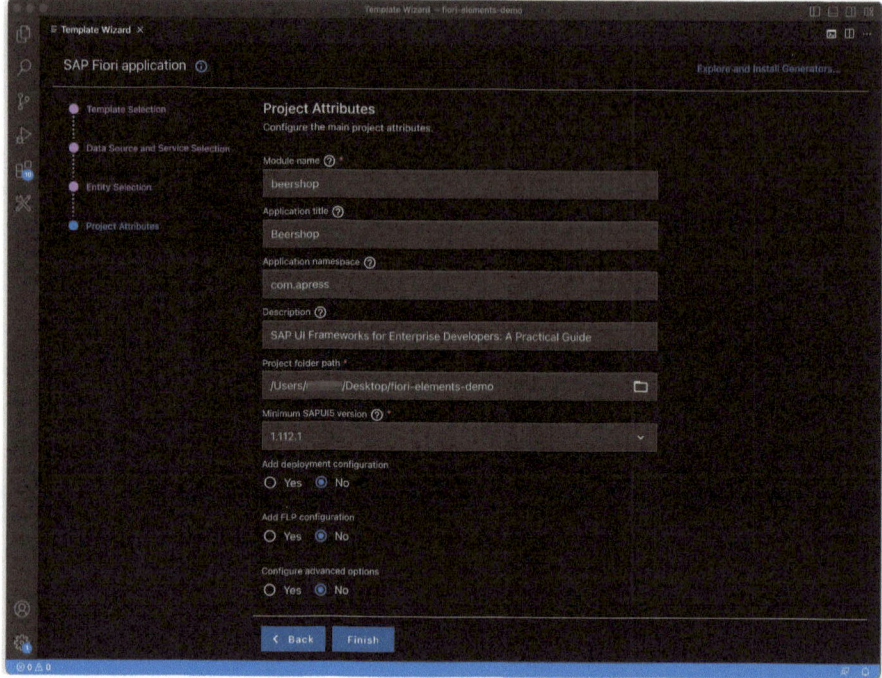

Figure 3-7. *The project attributes defined in the finishing step*

Give It a Test Run

Once all dependencies are installed, it's time to run the app. But to display data, we need to connect our local setup to a data source first. For development purposes, we can also leverage the mock data generator that works out of the box and is also part of the extension kit. To invoke it, you only need to run the following snippet:

```
cd beershop
npm run start-mock
```

You should see that the process is triggered, and, after a few seconds, a new browser tab should open. You'll notice that it takes a few seconds until you see the web app loading. This loading time is definitely longer than what you saw in the UI5 app of the previous chapter. But, to be fair, you get plenty of features for free, and the Fiori app also comes with relaxing loading animations. Now click **Go** to see the generated sample data (Figure 3-8). Admittedly, the mock data generator doesn't generate realistic data. It's OK if you only want to see some data to be able to test the filtering, sorting, or column rearrangement features of the list report application. You can even click a beer and see all its details on the object page.

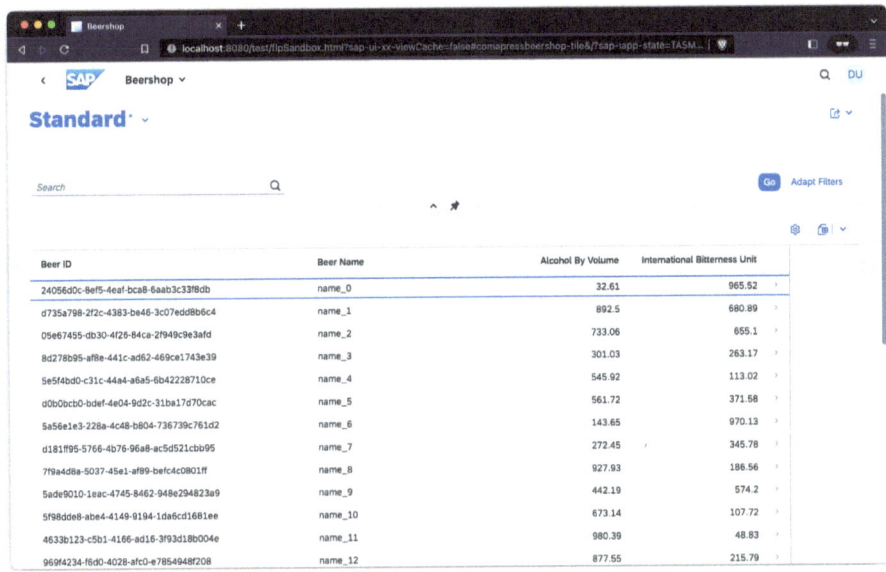

Figure 3-8. *The list report shows mock data*

Let's add some testing data that is more realistic. First, cancel the currently running npm run start-mock operation in your terminal.

Then, open the file ui5-mock.yml and inspect it. At the end of this file, you'll see a line saying generateMockData: true. **Remove this line** to skip that step. The file also contains a line saying mockdataPath: ./webapp/localService/data. This is where we need to add our sample data in json format. The following script will help you to download sample data before you restart the local development server. Run it in a new terminal session:

```
cd beershop/webapp/localService/
mkdir data
curl https://raw.githubusercontent.com/Apress/SAP-UI-
Frameworks-for-Enterprise-Developers-by-Marius-Obert-Volker-
Buzek/main/apps/fiori-elements/webapp/localService/data/Beers.
json > data/Beers.json
curl https://raw.githubusercontent.com/Apress/SAP-UI-
Frameworks-for-Enterprise-Developers-by-Marius-Obert-Volker-
Buzek/main/apps/fiori-elements/webapp/localService/data/
Breweries.json > data/Breweries.json
cd ../..
npm run start-mock
```

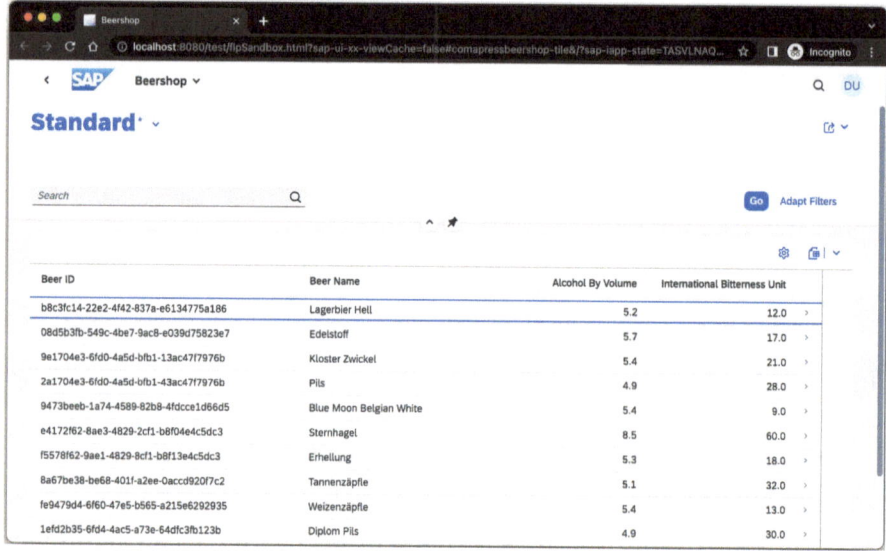

Figure 3-9. *The list report with sample data*

Note that `npm run start-mock` always opens a new browser tab.

This sample data looks much better now (Figure 3-9). Let's create a save point with git before we start to customize this application. This will help us to see the code changes that tooling does in the following steps. **Initialize a new git repository** via these terminal commands:

```
git init
git add .
git commit -m "initialized by the wizard"
```

Filter the Data

In an SAP Fiori list report application, the filter bar is located above the list and allows users to filter the data displayed in the list view by specific criteria. The filter bar typically includes several different filter fields, each representing a different attribute of the data being displayed. Users can apply filters by entering values into the filter fields. The filter bar also includes buttons to apply or clear filters or to open an advanced filter dialog with more options. The advanced filter dialog allows users to specify complex filter criteria using logical operators such as "and," "or," and "not." When the user applies a filter, the list view updates to only show the data that matches the specified criteria.

Developers can use annotations to define specific filter fields that will be available by default in the filter bar. These annotations can specify the data type, the UI type, and the label for the filter field. This allows developers to easily control which fields are easy to use for filtering and how they are presented in the filter bar.

We will now use the *Page Map* which is part of the *SAP Fiori tools* to make this change. Open the *Command Palette* and start typing Page Map so the palette reads ">Page Map" and select **Fiori: Show Page Map** as soon as the option is suggested. You should now see a map with two pages, list report and object page, and the property panel on the right-hand side. The page map does not only show the existing pages and their navigation paths

145

but also allows you to add new ones, configure them, or remove obsolete pages. For now, we want to modify the *List Page*. **Click the pencil icon** on the header of the list report box to open the *Page Editor* (Figure 3-10).

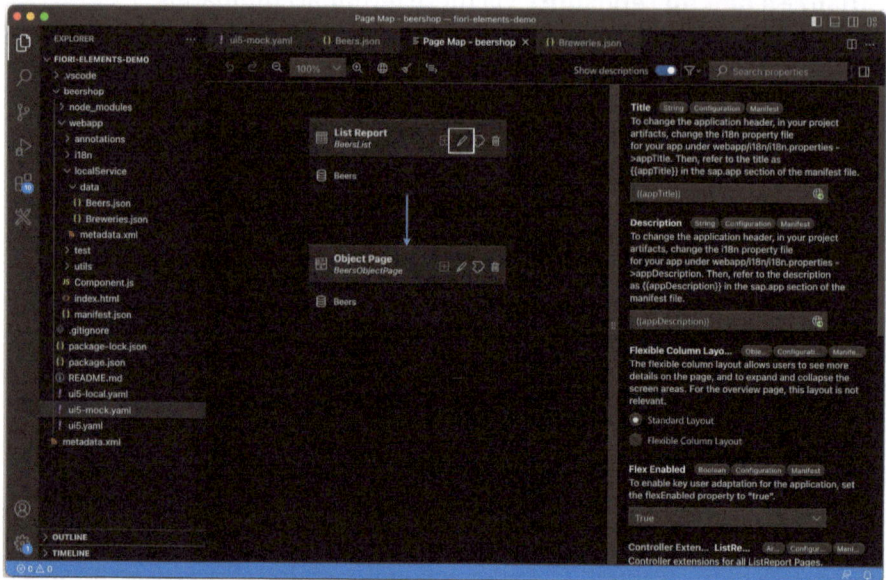

Figure 3-10. *The page map with the button to edit the list report page*

In the *Page Editor*, select the filter fields row to make the **Plus icon** visible. **Click** this button to **Add Filter Fields** and select the fields name, abv, and ibu in the popup. Confirm the popup with a **click on the Add button** (Figure 3-11).

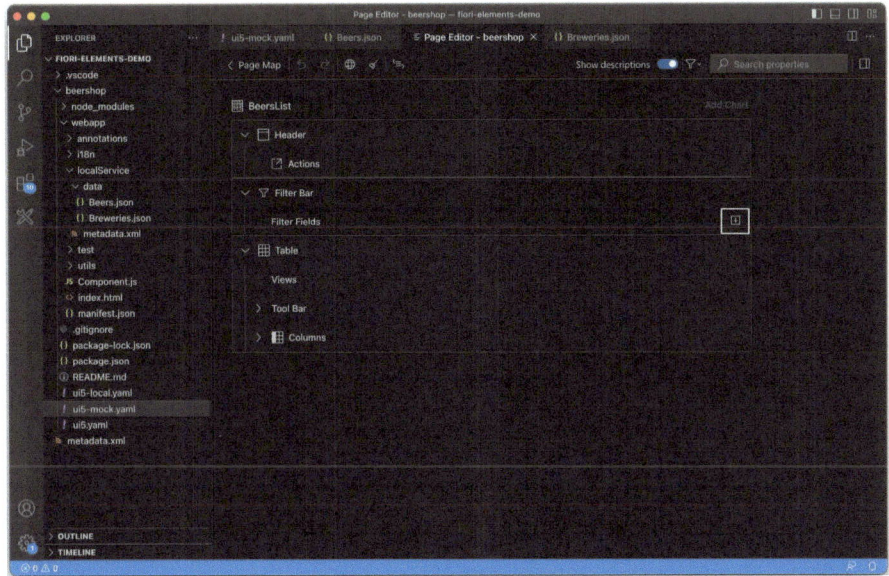

Figure 3-11. *The Page Editor with the button to add new filter fields*

Now refresh the still-running SAP Fiori elements application if not done automatically. You should now see two filter fields that are displayed by default. Let's say we want to see all beers that contain the substring "ell" in their name and have an alcohol by volume value of at least 5%. Enter ***ell*** and **>=5** in the new filter fields, respectively, and confirm with the **enter key**. There should be ideally only three beers left in the list (Figure 3-12).

If you don't see data showing up, try removing the `sap-iapp-state` URL parameter to clear the state and reload the page. Then, hit "Go" again.

147

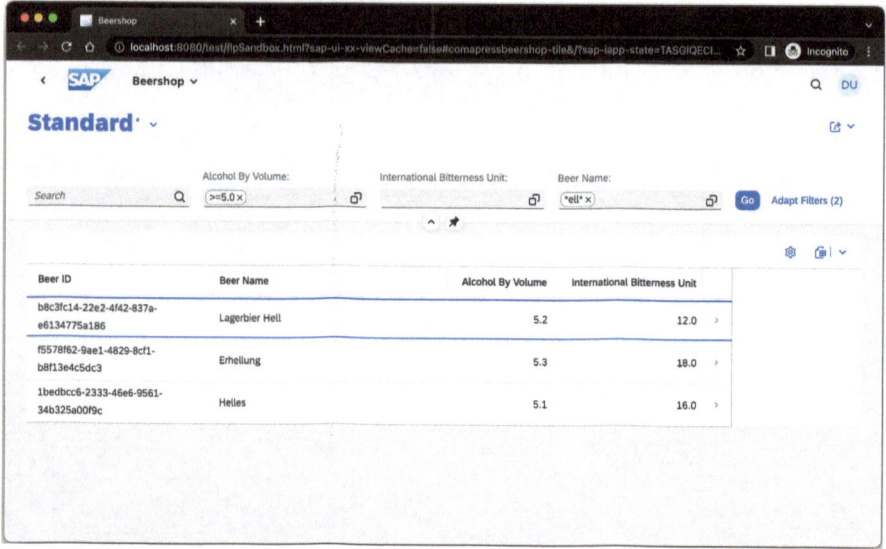

Figure 3-12. *The list report only shows the three beers that meet the filter criteria*

Hide Irrelevant Columns

The table in an SAP Fiori list report app is used to display data in a tabular format. The user can configure it to display specific columns and fields of data. Its main purpose is to present the data in a clear and easy-to-understand format to the end user. Too many columns can cause confusion on such pages and lead to negative performance. So it's up to the frontend developer to pick the right columns that are visible by default. These annotations can also be modified via the *Page Editor* of the list report. In case you already closed it, reopen it from the Command Palette of VS Code. The last row of the page box should be an expandable item that reads **Columns**. Some columns need to be added via the **Plus icon** and the option **Add Basic Columns**. The feature to rename columns can be found in the *Properties Panel* on the right under the section **Label** (Figure 3-13).

Reorder, rename, and add columns so that the list reads as follows:

1. Brewery

2. Beer

3. Alcohol By Volume

4. International Bitterness Unit

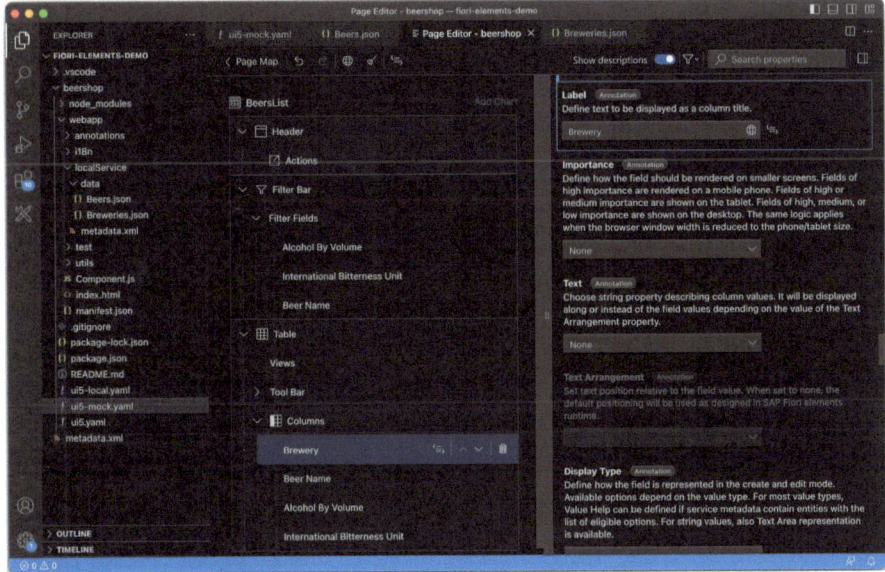

Figure 3-13. *The Page Editor shows the desired column order*

The *Page Editor* will save the changes automatically, so you can directly take a look at the changes in the browser when you reload the web app (Figure 3-14).

If you don't see data showing up, try removing the `sap-iapp-state` URL parameter to clear the state and reload the page. Then, hit "Go" again.

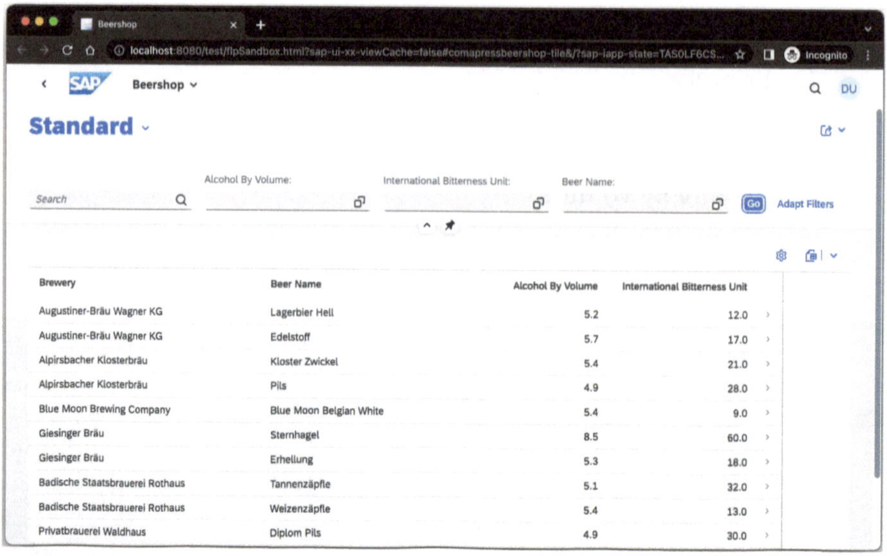

Figure 3-14. *The new table order*

Autoload Data

With SAP Fiori elements, you immediately benefit from a magnitude of features that make it easy to find the right data. One really convenient one is, for example, the infinitely scrolling that reloads new data when you scroll down the list. Many of these features can be toggled on or off, depending on whether you think it's useful for the app. A popular configuration is the auto-load flag. Turned off, the initial fetching of the data won't happen until the user enters clicks "Go." Turned on, it shows the first line items right away when the user accesses the page. You might guess that this feature toggle can be found in the *Properties Panel* of the table of the list re*port page*. And you could be right. But let's assume for a second you only know there is such an option, but you don't know where to find it. A *guided* approach is very helpful here because otherwise, you would need to search through many pages within the page map to find the right place. The better way is to open, once again, the **Command Palette** and search for "Fiori: Open Guided Development." The new user interface automatically selects your current project and provides a search bar in the top-right corner. **Type** Load in there, and you should see that the number of the available options is reduced to only one. This option also belongs to the list report page, which means you're on the right track. **Click this option** to see additional information in a panel and click the button that says **Start Guide** to get a detailed explanation about this implementation step and the possible values for the property Mode (Auto, Enabled, Disabled). Make sure **Enabled** is selected and click **Insert Snippet** to use this option in your project (Figure 3-15). A new file will open to confirm this, but there's nothing you need to do, and you can close it immediately again.

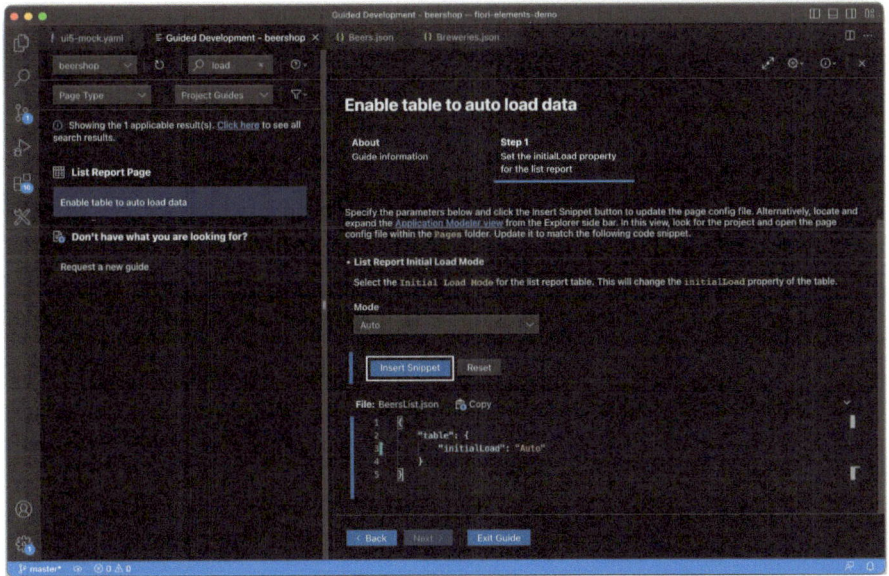

Figure 3-15. *Guided Development plugin showing the "Insert Snippet" button*

Go back to the browser and refresh the running web app. Now the data is visible right away without having to click **Go**.

A Header for Details

The object page is a layout that is used to display detailed information about a specific item or record of the list report and provides a more detailed view of the data associated with that item. These pages typically contain multiple sections, each displaying different types of information such as text, images, tables, and charts. It also usually contains a header section with information about the item's properties and a navigation bar that allows users to navigate between related items. Additionally, it also provides the functionality to perform actions on the selected item like Edit and Delete. It can even provide custom, entity-specific actions such as "Rate Beer" that trigger custom backend logic.

Select any of the beers to navigate to the object page. When you look at this page of our demo app, you'll mostly see a blank background besides a single "Details" section that contains all the information (Figure 3-16).

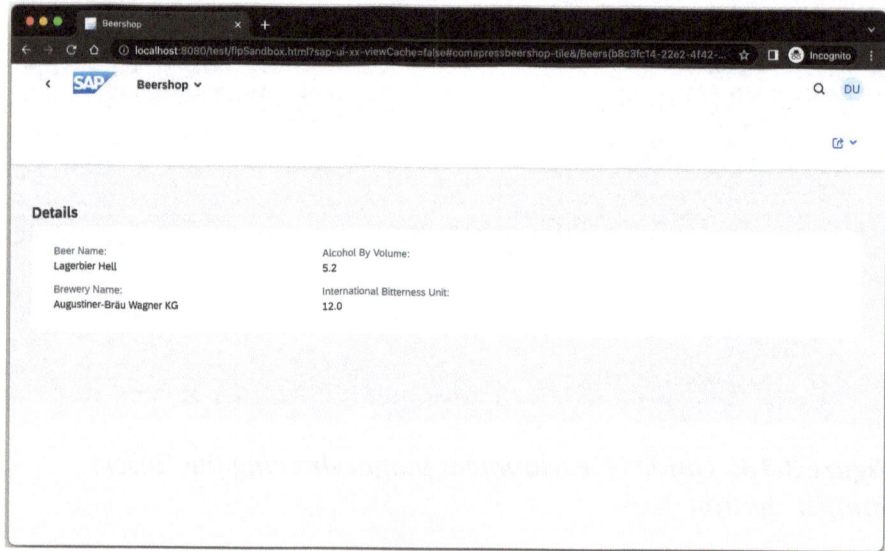

Figure 3-16. *A mostly empty object page*

Let's fill this page with life. Use the **Command Palette** to navigate to the *Page Map*. This time, you need to click the **pencil icon** on the object page. The *Page Editor* of this page will offer you to edit the properties of the header. Set the **Title to the name property** and **Description to the kind Property** and then choose the **brewery/name** for this particular property used for the description (Figure 3-17). To bring a bit more color to the empty screen, we'll also set the `image` to `brewery/logo`. This property might be empty in our mock data, but the object page will render a placeholder anyway as it knows our intention to display an image.

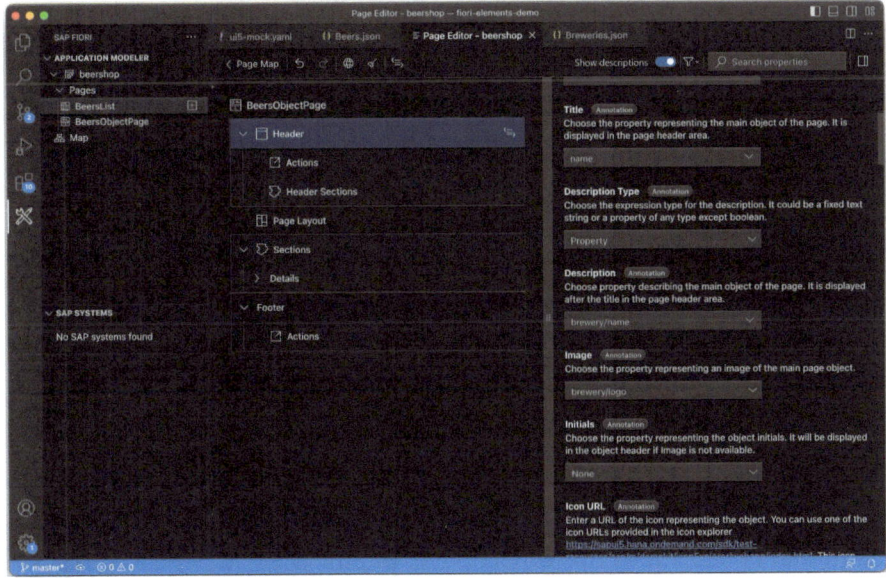

Figure 3-17. *The Page Editor configuration of the object page*

This Page Editor also allows you to add sections to the header that visualize the most important information of this object, which could be of interest to advanced users. For now, we're happy with the basic header that we just defined (Figure 3-18).

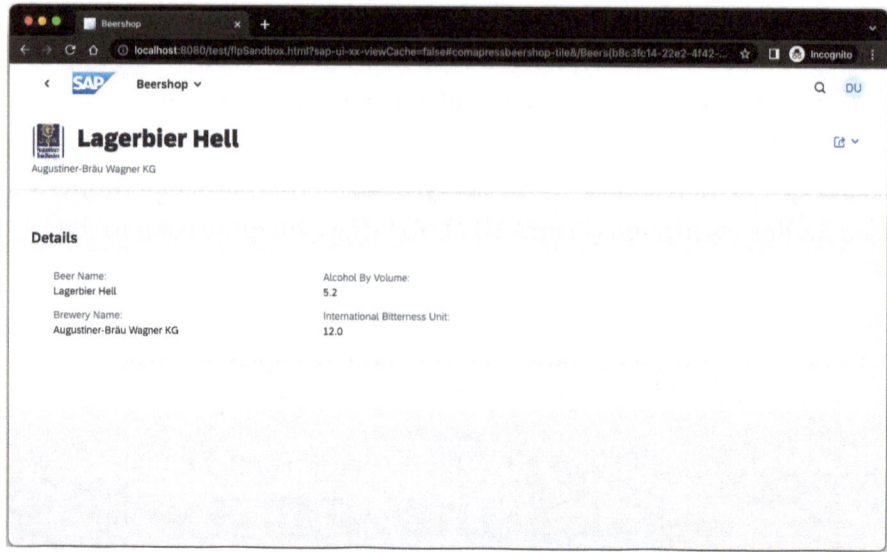

Figure 3-18. *The object page with header information from our sample data*

A Custom Fragment

The same Page Editor that allows you to add header elements can also be used to change the items that are displayed in the "Details" section and even to add and remove entire sections. This is also a good exercise for you if you want to dive deeper. Essentially, the sections will always follow the layout that is defined by this object page. But maybe you need to display additional information or functionality that is not provided by the default layout. A custom fragment that's literally a UI5 fragment is a reusable UI component that can be added to the object page to provide additional functionality or display additional information. Some common reasons to add a custom fragment include

- Displaying additional fields or data not provided by the default layout

- Adding new functionality such as custom buttons or actions

- Customizing the layout of the object page to better suit the needs of the application

- Providing a different view of the data depending on the item selected

Click the **Plus icon** of the **Sections row** in the Page Editor to see the different types of sections you can add. **Select "Add Custom Section"** here to open the popup (Figure 3-19).

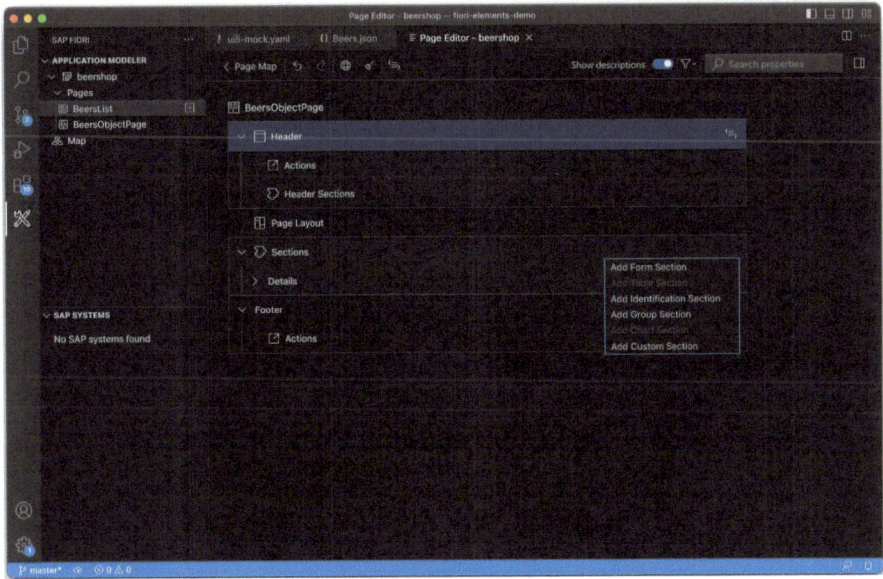

Figure 3-19. *The Page Editor offering multiple types of options*

This popup will ask for multiple properties of the new section like the name and the fragment and controller it will be based on. Complete this form as displayed in Figure 3-20 and confirm with **Add** to show it on the object page right under "Details."

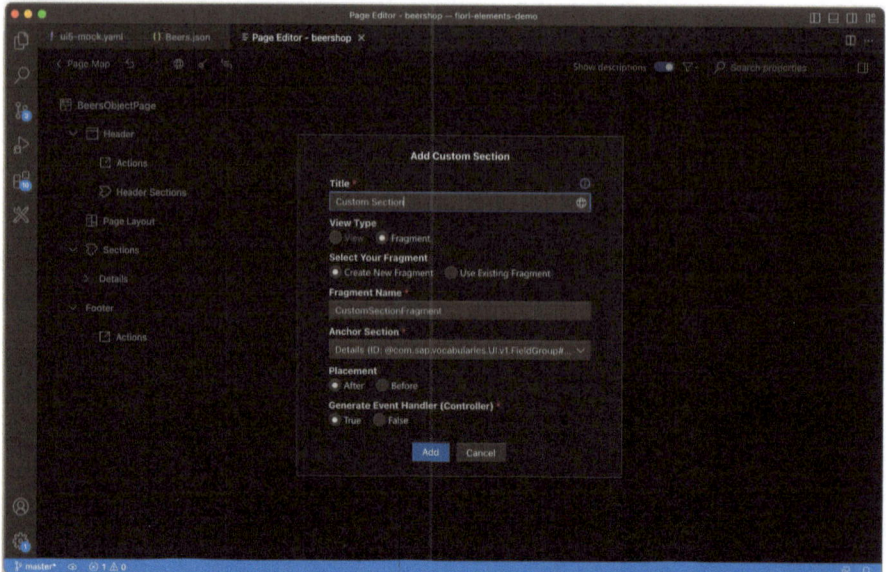

Figure 3-20. *Configuration to create a custom fragment in an object page*

Refresh the web application to see a new section that, so far, contains only a button. Click this button to see an info message hovering over the screen. This also makes sure that a new controller with the event handlers has been generated too (Figure 3-21).

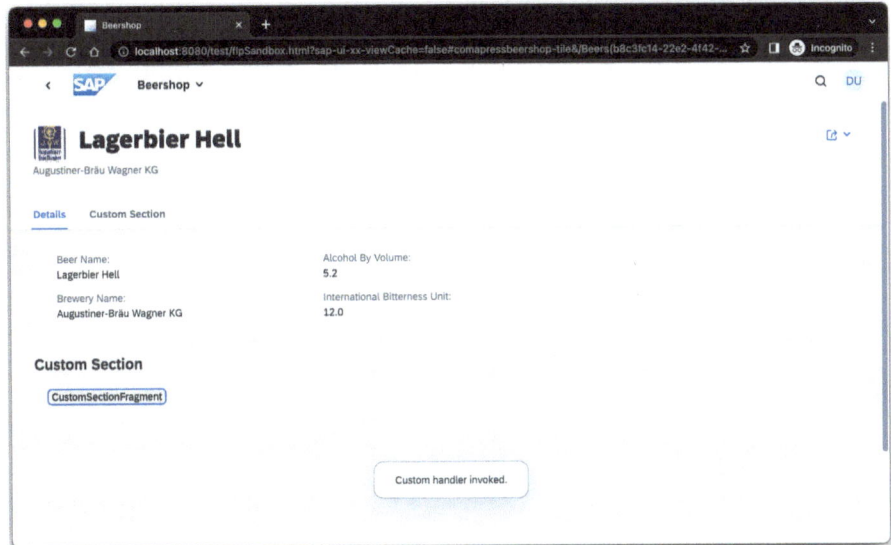

Figure 3-21. *The custom section after the button has been clicked*

This section is as of now pretty empty, but you can arbitrarily extend the webapp/ext/fragment/CustomSectionFragment.fragment.xml and webapp/ext/fragment/CustomSectionFragment.js files with your knowledge from the previous chapter.

Inspecting the Changes

The demo application that we built so far deviates in a couple of places from the standard app that we generated initially. For all of these changes, we had to write very little code – most of the building process was done by the SAP Fiori tools. Git and the source control view[13] of VS Code (Figure 3-22) help us to see these modifications in the project source code.

---

[13] https://code.visualstudio.com/docs/introvideos/versioncontrol

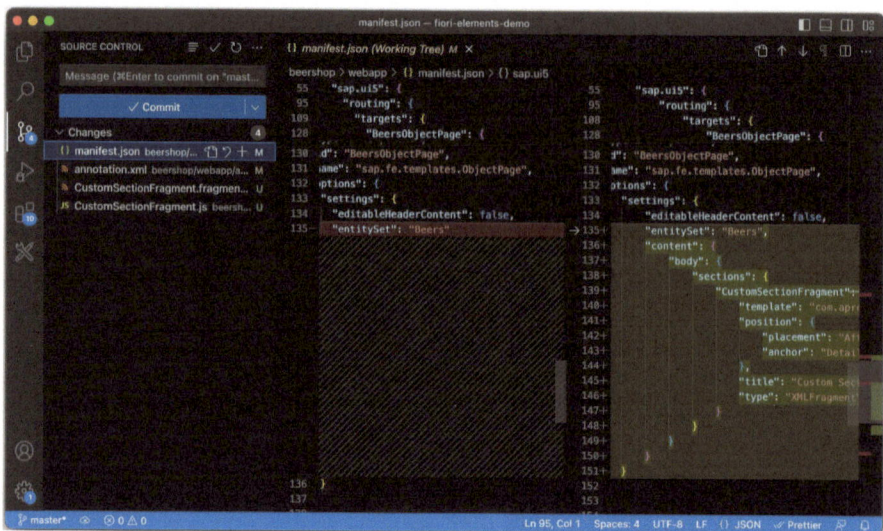

Figure 3-22. *The diff of the files that were modified in this section*

The annotations.xml file is used to define the metadata for the application. It contains information such as the entities and fields that are used in the application, the relationships between the entities, and the behavior of the application when it is launched. The Fiori tools modify this file that overrides the properties of the OData service to ensure that the SAP Fiori floorplans look and behave the way the developer intended.

Another file they modified frequently is the manifest.json file, which is the configuration file that contains the settings and metadata for the application. It includes information such as the application's name, the routes for navigation, the theme and icons, and the required libraries and dependencies. The Fiori tools modify this file that you already know from normal UI5 apps to ensure that the application is configured correctly and can be launched and accessed by users.

Other files you already know are the fragments, written in XML, and the corresponding controllers, written in JavaScript. These files were also generated by the Fiori tools and could now be implemented in the same way we learned in the previous chapter.

Make the App Robust

Just as with UI5 in the previous chapter,[14] the obvious thing to do to make the SAP Fiori elements application more robust is to write tests for it. However, given that most UI parts are auto-generated, the challenge is to come up with tests that don't test the framework (and its auto-generating logic), but your custom app configurations and implementations.

Subsequently, writing Unit tests for SAP Fiori elements applications only makes limited sense: most (if not all) of the app logic is declared via annotations and dynamically generated by the framework. There is no point in unit testing this, as it would mean testing framework functionality. Instead, we recommend focusing on writing Integration and End-to-End tests, with OPA5 and wdi5, respectively.

In the case of our sample app, testing code/configuration under our responsibility applies to the auto data load at app start. By default, the initial view (list report) doesn't show data. But we configured the app to start querying the beers immediately, not requiring any user interaction. A sample test case would be to assert that the immediate navigation to the detail page of a beer is possible, without having to trigger any prior user interaction.

With the auto-generated parts of SAP Fiori elements feeling like magic happening at runtime, there is fortunately also a test library providing complementary actions to the "magic": the OData v4 test library (referred to as "test library" in short). It provides test functions for many of the baked-in capabilities of an SAP Fiori elements app. For example, the default navigation possibility from the list report page to the detail page can be tested with a prebuilt `.onTable().iPressRow($i)` method of the test library.

---

[14] Chapter 2, "Dual: SAPUI5 and OpenUI5".

The OData v4 test library is available for both OPA5[15] and wdi5[16]; its API documentation is at `https://ui5.sap.com/#/api/sap.fe.test`. Even better, the commands and operations of the test library are API-aligned in OPA5 and wdi5. With this, the notation is the same with both frameworks (Listing 3-5).

Listing 3-5. Same test library API in OPA5 and wdi5

```
When.onTheBeersList.onTable().iPressRow(0);
Then.onTheBeersObjectPage.iSeeThisPage();
```

OData v4 Test Library with OPA5

For OPA5, the Fiori tools generator already provided a skeleton file layout. To bootstrap the test library, the `sap.fe.test.JourneyRunner` needs to be required and programmatically configured. Find an example in our repo at `apps/fiori-elements/webapp/test/integration/opaTests.qunit.js` (Listing 3-6).

Listing 3-6. Using the OData v4 test library in OPA5

```
sap.ui.require([
    "sap/fe/test/JourneyRunner",
    //...
  ],
  function (JourneyRunner, /* ... */ ) {
    var Runner = new JourneyRunner({
        launchUrl: sap.ui.require.toUrl("com/apress/beershop")
        + "/index.html",
    });
```

---

[15] https://blogs.sap.com/2022/10/03/opa5-test-library-for-fiori-elements-v4-applications/

[16] https://ui5-community.github.io/wdi5/#/fe-testlib

```
Runner.run({
    pages: {  /* ... */  },
  },
  opaJourney.run
);
});
```

From there on, it is a matter of using the proper page layout patterns (e.g., `sap.fe.test.ListReport`) with predefined operations of that page kind (e.g., for the ListReport `onTable().iChangeSortOrder(...)`[17]).

At this point, reading code is more helpful than displaying abbreviated fragments, so head over to our sample repo and look through the OPA5 files using the OData v4 test library at `apps/fiori-elements/webapp/test/integration/**/*`!

Running the test library–enhanced OPA5 tests happens thanks to the SAP Fiori tools.[18] It is an extended version of the UI5 Tooling maintained by the SAP Fiori elements team and exposes a `run` function similar to `ui5 serve`. Combined with using the `sap-fe-mockserver` for providing mock data at runtime in a `ui5-mock.yaml` (see `apps/fiori-elements/ui5-mock.yaml` in the sample repo), the tests can be run with `npx fiori run --config ./ui5-mock.yaml` and opening `http://localhost:8081/test/integration/opaTests.qunit.html` in your browser.

Hint: If port 8081 is already used by another application, the tooling will choose the next available port and display it at startup: `URL: http://localhost:xxxx`.

---

[17] https://ui5.sap.com/#/api/sap.fe.test.api.TableActions%23methods/iChangeSortOrder
[18] www.npmjs.com/package/@sap/ux-ui5-tooling

OData v4 Test Library with wdi5

For using the test library with wdi5, the first step is the general installation of wdi5 – most comfortably accomplished via npm init wdi5@latest in the root directory of your SAP Fiori elements app.

Then just as the documentation at https://ui5-community.github. io/wdi5/#/fe-testlib explains, bootstrap the test library as a façade of browser.fe in your test file (see apps/fiori-elements/webapp/test/e2e/ FirstJourney.test.js in the sample repo) as shown in Listing 3-7.

Listing 3-7. Bootstrap and use the OData v4 test library in wdi5

```
describe("v4 test lib", async () => {
  let FioriElementsFacade
  before(async () => {
    FioriElementsFacade = await browser.fe.initialize({
      onTheBeersList: {
        ListReport: {
          appId: "com.apress.beershop",
          componentId: "BeersList",
          entitySet: "Beers"
        }
      /* ... */
    })
  })
  // the actual tests
  it("...", async () => {
    await FioriElementsFacade.execute((Given, When, Then) => {
      // ...
      When.onTheBeersList.onTable().iPressRow(0)
      // ...
    })
  })
})
```

The `browser.fe.initialize` expects the same configuration of page layouts (e.g., `sap.fe.test.ListReport`) as in OPA5. Subsequently, the configured page's API methods can be used in the actual it tests (here: `onTable().iPressRow(0)` in Listing 3-7).

Find a comprehensive example of using wdi5 with the OData v4 test library not only in wdi5's repo at `https://github.com/ui5-community/wdi5/blob/main/examples/fe-app/webapp/wdi5-test/Journey.test.js` but also in this book's repo at `apps/fiori-elements/webapp/test/e2e/**/*`.

Running wdi5 with the OData v4 test library is no different from running wdi5 itself:

1. In one terminal, start up the application.

    Here: `npm run start-mock` (for running things with mock data)

2. In another terminal session, run wdi5.

    Here: `npm run wdi5`

For running wdi5 with test library commands in different browsers, no specific test library config is needed. Instead, the same configuration options as with wdi5 (and wdio)[19] itself apply!

CI and the OData v4 Test Library

When using the test library, the configuration for both OPA5 and wdi5 is no different from running any other integration or end-to-end test setup as explained in Chapter 2; we already took a closer look at running test scopes in CI.[20] Also, all concepts and recommendations mentioned in that chapter for running tests in a CI environment also apply as to the OData v4 test library–based tests.

---

[19] `https://ui5-community.github.io/wdi5/#/configuration`
[20] Chapter 2, "Make the App Robust".

In the case of OPA5, `karma` is used as the test runner, looking for the top-level `apps/fiori-elements/webapp/test/testsuite.qunit.html` that wraps all integration tests. Most CI environments don't have a graphical environment on the machines executing the tests. For this reason, the karma config uses headless Chrome for running the tests in `apps/fiori-elements/karma-ci.conf.js`.

Similarly with wdi5, the `--headless` flag in `apps/fiori-elements/package.json` is forwarded to `apps/fiori-elements/wdio.conf.js`, telling the framework to run Chrome in headless mode.

To execute the tests, switch to the root folder of the app, `apps/fiori-elements/`. For the integration scope, run `npm run ci-test`. For the end-to-end scope, first start the app via `npm run start-mock`, then in another terminal `npm run wdi5`.

In both cases, you'll see the tests running in headless mode.

Deploy Continuously

For an automated deployment of an SAP Fiori elements app, the same principles apply as for a freestyle UI5 app: make automated tests a prerequisite and first cut the test, build, and publish process locally. Then transfer the entire flow to a central DevOps platform. If the build and deploy process is triggered by a CI or some other development event, then it is commonly referred to as "deployed continuously."

The deployment target for an SAP Fiori elements app most likely differs from a UI5 freestyle app. The UX and design options of SAP Fiori elements UIs are standardized and work best in a launchpad environment. The latter is a UX floorplan itself[21] and describes itself as "(…) the launchpad displays a home page with tiles. Each tile represents a business application

---

[21] https://experience.sap.com/fiori-design-web/launchpad/

that the user can launch. Tiles can also display live status indicators, such as the number of open tasks. The launchpad is role-based. In other words, the user's role determines which app tiles are shown." The launchpad UX pattern is productized by SAP as the "SAP Fiori launchpad on-premise[22]" and "SAP Build Work Zone.[23]" Both represent the "best fit" deployment target for SAP Fiori elements apps.

In our sample app repository, we have provided a build script that prepares our SAP Fiori elements sample for deployment. First, change into the proper directory via `cd apps/fiori-elements`. Then start the build with `npm run build:wz` ("wz" stands for "Work Zone"). The build step uses the UI5 Tooling task extension "ui5-task-zipper"[24] to compress the tooling's standard build result into a zip file. This archive can then be used for subsequent deployment into your target environment. Marius has written a detailed blog post[25] on that topic.

Our sample app only uses a mock server and no real backend. If deployed directly, it will not work out of the box. Look at the app's `manifest.json` (apps/fiori-elements/webapp/manifest.json) setting for the backend URI. Change `"uri": "/here/goes/your/serviceurl/"` to plug in a real backend for the app.

---

[22] https://help.sap.com/docs/SAP_FIORI_LAUNCHPAD?version=EXTERNAL&locale=en-US

[23] www.sap.com/products/technology-platform/workzone.html

[24] https://github.com/ui5-community/ui5-ecosystem-showcase/tree/main/packages/ui5-task-zipper

[25] https://blogs.sap.com/2020/10/02/serverless-sap-fiori-apps-in-sap-cloud-platform/

While a launchpad incarnation is a proper deployment target, there's also an equivalent for the dev time. Automatically generated by the Fiori tools when bootstrapping an SAP Fiori elements app, the file `apps/fiori-elements/webapp/test/flpSandbox.html` mimics a launchpad UX pattern. It can be used to test the app under development in an isolated sandbox similar to the deployment target (Figure 3-23).

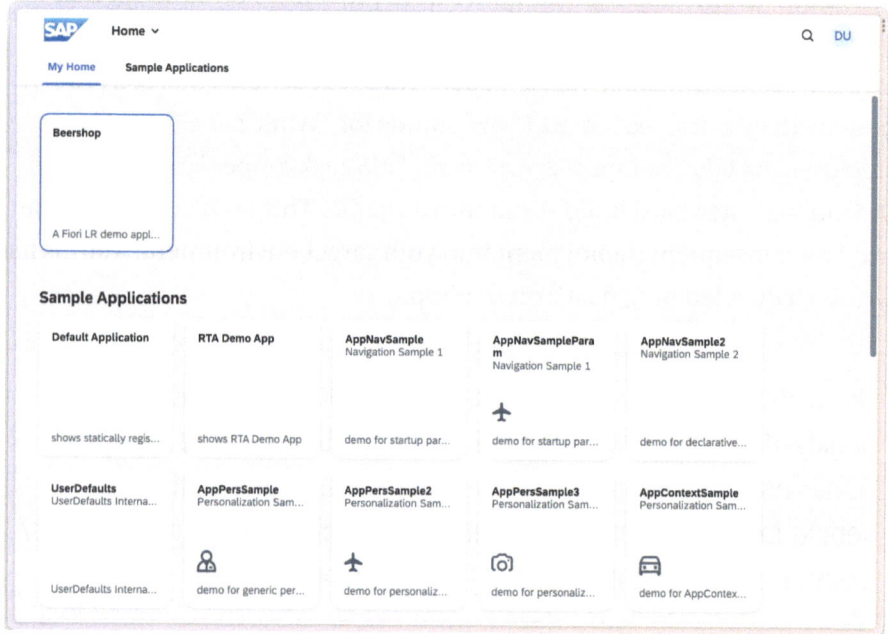

Figure 3-23. *`flpSandbox.html` mimics a launchpad UX pattern at dev time*

The sandbox launchpad is automatically started with `npm run start-mock` in the `apps/fiori-elements` folder.

Include Parts of SAP Fiori Elements in UI5 Freestyle Apps

When custom UI5 code is included in an SAP Fiori elements app, it is referred to as a "breakout" – in the sense of breaking out of the boundaries imposed by the standard UI5 controls available as SAP Fiori elements UI parts.

The other way around, when SAP Fiori elements pieces are put into a UI5 freestyle app, those are considered Building Blocks.[26] Building Blocks are reusable artifacts of UI controls and logic from the SAP Fiori elements library. When presented metadata and backend data at runtime, they auto-generate UI elements with interaction logic. And because they are living in the sap.fe.macros namespace, they are also called macros. Both breakouts and Building Blocks comprise large parts of the SAP Fiori elements flexible programming model.[27]

There's a step-by-step guide on how to enhance each view of a UI5 freestyle app to work with Building Blocks at Flexible Programming Model Explorer.[28] Keep in mind that as a prerequisite, an OData v4 service is needed that already holds proprietary SAP Fiori elements annotations from the com.sap.vocabularies namespace (see the "SAP Fiori Elements Configuration" section earlier in this chapter). Only with those annotations in place can the Building Blocks be utilized in custom apps.

---

[26] https://ui5.sap.com/test-resources/sap/fe/core/fpmExplorer/index.html#/buildingBlocks/buildingBlockOverview

[27] https://ui5.sap.com/test-resources/sap/fe/core/fpmExplorer/index.html#/overview/introduction

[28] https://ui5.sap.com/test-resources/sap/fe/core/fpmExplorer/index.html#/buildingBlocks/guidance/guidanceCustomApps

Additionally, bootstrap an SAP Fiori elements–enhanced freestyle app only from the SAPUI5 CDN and not with the UI5 Tooling or the OpenUI5 CDN. The npm module holding the required `sap.fe` sources for the UI5 Tooling is restricted for SAP internal use only[29] (Listing 3-8).

Listing 3-8. Bootstrap the UI5 app with Building Blocks from CDN

```
<script
  id="sap-ui-bootstrap"
  src="https://ui5.sap.com/resources/sap-ui-core.js"
  ...
></script>
```

Also, the `sap.fe.macros` are aggressively cached. So at dev time, changes related to a Building Blocks tag don't update on reload per default. They would require a manual deletion of cache artifacts in the development tools of the browser. However, caching of Building Blocks can be turned off by setting `sap-ui-xx-viewCache` to `false` in the UI5 bootstrap (Listing 3-9).

Listing 3-9. Turn off caching of Building Blocks with `xx-viewCache` set to false

```
<script
  id="sap-ui-bootstrap"
  src="https://ui5.sap.com/resources/sap-ui-core.js"
  data-sap-ui-theme="sap_horizon"
  data-sap-ui-resourceroots='{
    "com.apress.openui5": "./"
  }'
  data-sap-ui-oninit="module:sap/ui/core/ComponentSupport"
  data-sap-ui-compatVersion="edge"
```

---

[29] www.npmjs.com/package/@sapui5/sap.fe

```
data-sap-ui-async="true"
data-sap-ui-xx-viewCache="false"
></script>
```

When working with SAP Fiori elements–enhanced UI5 apps, it is also recommended to restart the UI5 Tooling (in its role as a web server) when changing local metadata annotations. Those don't get picked up on browser reload alone, but need a new ui5 serve to induce UI5 changes.

The Road from Here

This hands-on section did show how it is possible to write such a list report application with almost no self-written code and, therefore, minimal maintenance obligations. This is also the reason why this highly standardized, yet customizable, app took only a couple of hours to build. Still, the demo application can handle 100s of beers while offering an intuitive and proven user experience to the end users. For real-life projects, this will of course, depending on the scope, take a bit longer and require multiple developers.

You also saw how the SAP Fiori tools extension for VS Code helps you during development. One common requirement for SAP Fiori elements, especially list report apps with many configuration options, is that they should come with some support to persist these configurations. With this, business users could save their filter and sorting settings in the list report to have easy access to their personal variant of the data. Similarly, they might not be interested in all sections of an object page and desire to hide some or change the labels of some fields. The UI5 Flexibility Services[30] do exactly this and can be practical if you are looking for such a service and want to embed the application in an SAP Fiori launchpad anyway.

---

[30] https://sapui5.hana.ondemand.com/sdk/#/topic/a8e55aa2f8bc4127923 b20685a6d1621.html

If you would like to learn more about this rapid development system, we recommend exploring the SAP Fiori design system.[31] With this knowledge, you can start integrating other SAP Fiori elements floorplans, like the overview page or a worklist, in this application and connect all pages via navigation paths in the *Page Map* of the Fiori tools. Or get to know the other features of the extension pack, such as the *Service Modeler*,[32] to visualize and refine the OData services your apps consume. Learning more about the annotation vocabulary might also come in handy to be able to infuse more information into these OData services. Did you know that these annotations can exist in multiple natural languages to ensure SAP Fiori elements apps can be localized?[33] The SAP Cloud Application Programming Model[34] and the ABAP RESTful Application Programming Model (RAP)[35] are probably the best ways to continue this learning path while writing scalable cloud services at the same time.

SAP Fiori Elements for Enterprise Applications

As SAP Fiori elements apps are technically just UI5 apps, many of the things we discussed in the previous chapter about enterprise readiness, such as security features and compatibility, also apply here. However, the additions that come with SAP Fiori elements also move the needle on some of these points. The palette of enterprise UI elements that can be used in applications became even larger with the floorplans and

---

[31] https://experience.sap.com/fiori-design-web/

[32] https://marketplace.visualstudio.com/items?itemName=SAPSE.sap-ux-service-modeler-extension

[33] https://cap.cloud.sap/docs/guides/i18n

[34] https://cap.cloud.sap/docs/

[35] https://community.sap.com/topics/abap/rap

the building blocks. This also has a positive impact with regard to the scalability and maintainability of entire applications. And it can also be a significant advantage for companies when they need to ship new solutions fast. These larger UI elements also offer to be highly extensible, as custom code fragments can be injected virtually anywhere on the floorplan that offers a magnitude of configuration parameters.

But, these advantages come with a price. There is no such thing as free lunch![36] The reusable UI blocks cannot be customized beyond what the configuration options offer. Extending the features would become very costly as many parts of the application are black boxes. Another disadvantage is the performance. If you implemented the UI5 app and now this SAP Fiori elements app, you probably noticed the UI5-only app loaded faster, even when the SAP Fiori elements apps try to reduce the perceived loading time with nice animations. The complexity that we didn't have to implement, still, had to be implemented by someone. This logic needs to be loaded and executed by the client. And chances are high that code that you don't need for your application will be loaded and executed – and when using SAP Fiori elements, you need to wait for this to happen. This is especially true for UI5 freestyle apps enhanced by Building Blocks. There's always an overhead of preloaded code outside the control of the developer, resulting in slower startup times. It's worth mentioning that the performance in an SAP Fiori launchpad environment is probably better than during development because multiple shared resources will be preloaded.

So, what can you do if you want to build an enterprise application that lives on the *opposite side of the scale*? An application that loads blazingly fast and gives you a lot of freedom, and the responsibility, to implement every little aspect of the frontend while still adhering to the SAP Fiori design system? Then the technology covered in the next chapter might be for you.

---

[36] https://en.wikipedia.org/wiki/There_ain%27t_no_such_thing_as_a_free_lunch

CHAPTER 4

Standard: UI5 Web Components

In this chapter, we're going to dive into the world of Web Components and explore the remarkable open source project UI5 Web Components. This project utilizes the Web Components standard to provide reusable, SAP Fiori conform components that can seamlessly integrate into any web application.

To kick-start your journey, we'll guide you through setting up your workspace, ensuring you have all the necessary tools at your disposal. We'll also provide a hands-on tutorial on how to bootstrap a project using UI5 Web Components without relying on any web framework.

We will then demonstrate how this groundbreaking technology can be embedded in existing web frameworks like UI5 and React. By harnessing the power of UI5 Web Components, you'll unlock the potential for simplified UI development, enhanced code reusability, and accelerated prototyping.

To wrap things up, we'll reflect on the potential of UI5 Web Components for enterprise applications and explore how they can streamline UI development, promote consistency, and facilitate faster prototyping, ultimately empowering you to create scalable and maintainable frontend applications that cater to the diverse needs of modern enterprises.

Ready? Let's get started.

© Marius Obert and Volker Buzek 2023
M. Obert and V. Buzek, *SAP UI Frameworks for Enterprise Developers*,
https://doi.org/10.1007/978-1-4842-9535-9_4

What Are Web Components?

Web Components[1] are a set of web standards that allow developers to create reusable and encapsulated custom elements. They are based on HTML, CSS, and JavaScript and can be used just like native HTML elements such as <button>. Each component has its own **style and behavior**, isolated from the rest of the page, a.k.a. in the Shadow DOM.[2] These custom elements are interoperable and can be shared across projects, making it easier to build and reuse UI components. Standardized features of modern browsers made this technology possible. Therefore, Web Components don't require a JavaScript framework at all to run. However, developers can combine them with any framework out there to make their lives easier.

SAP used this technology to build a catalog of controls that were previously only available in UI5, in a more modular and flexible way: the UI5 Web Components.[3] The name includes "Web Components" to refer to the used standards. While the UI5 prefix might suggest otherwise, **UI5 Web Components is neither built with SAPUI5/OpenUI5 nor is it its successor**. However, it brings UI5 qualities and the SAP Fiori UX to the HTML level, so it can be used with any web framework that supports Web Components.

This addresses the wishes many developers in the SAP ecosystem had for a long time. They often criticized the core concepts of UI5, like the MVC pattern, and claimed that newer web technologies would provide a more performant foundation for their apps. SAP listened to these critics and created UI5 Web Components – and made this technology open source. By using UI5 Web Components, developers can take advantage of skills

---

[1] www.webcomponents.org/

[2] https://developer.mozilla.org/en-US/docs/Web/Web_Components/Using_shadow_DOM

[3] https://github.com/SAP/ui5-webcomponents

and technologies that were previously not relevant in the SAP UI universe. They can now bring their favorite web development tools and technologies to work. This freedom has the potential to offer several advantages over traditional approaches to SAP web development. But it also comes with more responsibilities.

The amount of available controls is still limited compared to UI5. This doesn't mean developers using UI5 Web Components are left with nothing. The most basic controls, such as buttons, labels, input fields, etc., are included (Figure 4-1). What is intentionally missing are mostly screen elements that require more space. Neither do UI5 Web Components prescribe a programming model with data binding, models, routers, or i18n features. And since there are no models, there are also no OData models. The technology is, in fact, unaware of the OData protocol, which means it's up to the developer to bring their own tools or do the footwork to implement these features when needed. We often see that Web Components are used for projects that connect to other HTTP-based protocols, such as REST or GraphQL, and therefore, they wouldn't need the OData features that come with UI5. Alternatively, UI5 Web Components can be combined with libraries that provide a programming model with the desired features. We'll cover this in the section "Usage with SPA Frameworks," but for now, let's start using the vanilla version.

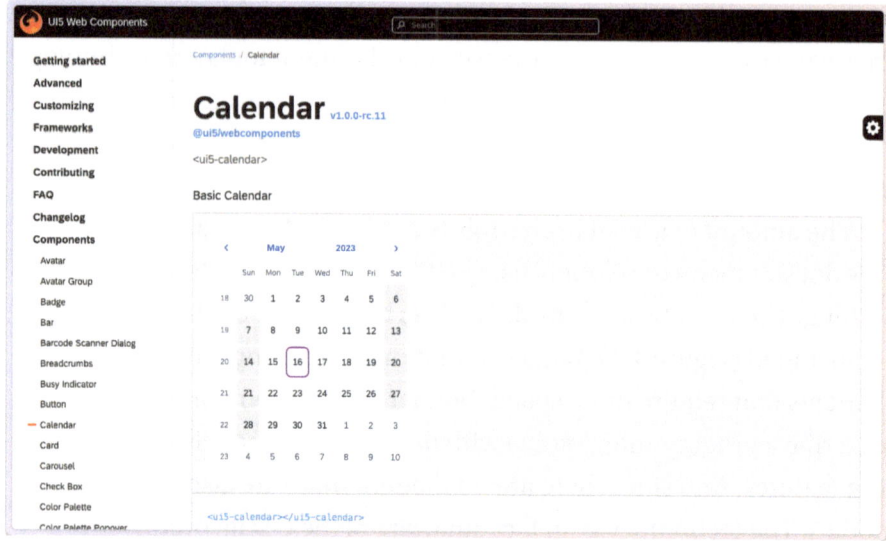

Figure 4-1. *The calendar[4] component of the playground page*

Setting Up Your Workspace

If you have completed the exercises in the previous chapters, you already have VS Code[5] and Node.js[6] installed. In case you skipped these chapters, please install these tools now.

Besides these tools, you should know Vite[7] (French for "fast"), but there's no need to install it. Vite is a build tool and development server for modern web projects, similar to the UI5 Tooling for UI5-based projects. It was created to improve the development experience by providing a fast and efficient way to build and serve web applications. Vite is designed to work with modern web technologies like EcmaScript modules, which enables faster and more efficient development and build times.

---

[4] https://sap.github.io/ui5-webcomponents/playground/components/Calendar/
[5] https://code.visualstudio.com/download
[6] https://nodejs.org/en/download/
[7] https://vitejs.dev/

It uses a dev server that watches your project files and serves them through a web server, allowing you to see the changes you make in real time as you code. This eliminates the need to manually refresh your browser and accelerates the development process.

Learning by Doing

In this exercise, we will build a web application with UI5 Web Components and avoid the usage of any Single Page Application (SPA) framework. While this is probably not really applicable to real-world use cases, it is a great way to learn which features are provided by UI5 Web Components and to understand why it makes sense to combine them with frameworks to boost efficiency.

Our demo application will consist of a shell bar, a data entry form, and a list that all follow the SAP Fiori design system *loosely*. Later in this section, we'll explain in more detail what we mean by "loosely." The form will consist of three inputs and a button for submission, while the list will display the same sample data that we already used in the previous exercise.

Bootstrap the Project

To start, use npm and Vite to bootstrap a new project (Listing 4-1).

Listing 4-1. Vite bootstrap

```
npm init vite
```

When prompted for a project name, use ui5-webcomponents and make sure to select Vanilla for the framework and JavaScript for the variant.

This command will create the structure of the project. Now, run these commands to install the initial dependencies and start a dev server (Listing 4-2).

Listing 4-2. Vite project start

```
cd ui5-webcomponents
npm install
npm run dev
```

You will see a message that a server is running on localhost. Open this URL in the browser of your choice. You'll see that Vite generated a page with some content for us. To start from a spotless plate, open VS Code and remove the following files:

- *public/vite.svg*
- counter.js
- javascript.svg

Override the content of main.js with Listing 4-3.

Listing 4-3. Custom style import

```
import "./style.css";
```

And **styles.css** should only contain these two selectors: **body** and **#page** (Listing 4-4).

Listing 4-4. Minimal CSS

```
body {
 margin: 0;
 min-width: 320px;
 min-height: 100vh;
}

#page {
 margin: 0 auto;
 max-width: 95%;
}
```

And finally, **add your first UI5 Web Component to the `index.html`** and remove all other nodes. Optionally, you can also change the favicon and title as shown in Listing 4-5.

Listing 4-5. UI5 Web Components vanilla usage

```html
<!DOCTYPE html>
<html lang="en">
  <head>
    <meta charset="UTF-8" />
    <link rel="icon" href="https://sap.github.io/
    ui5-webcomponents/assets/images/favicon.ico"
    type="image/x-icon">
    <meta name="viewport" content="width=device-width, initial-
    scale=1.0" />
    <title>SAP UI Frameworks for Enterprise Developers:
    A Practical Guide</title>
  </head>
  <body>
    <ui5-button>Hello UI5 Web Components</ui5-button>
    <script type="module" src="/main.js"></script>
  </body>
</html>
```

Your browser page will be automatically refreshed, and you should now see (Figure 4-2) a page saying "Hello UI5 Web Components" – but not rendered as a button. This is expected as our project is not yet aware of UI5 Web Components. Let's change that!

Hello UI5 Web Components

Figure 4-2. *An empty page with no SAP Fiori content*

Install the UI5 Web Components Dependencies

The UI5 Web Components are split into two modules: @ui5/
webcomponents and @ui5/webcomponents-fiori.

The first module can be seen as the "main" package for general web
apps and contains components such as buttons, pickers, inputs, list, table,
common icons, etc. The second module @ui5/webcomponents-fiori
provides more semantic, higher-order components that implement SAP
Fiori patterns and additional icons.

Run the command in Listing 4-6 in a new terminal session to install
both modules to the project.

Listing 4-6. Install UI5 Web Components

```
npm add @ui5/webcomponents @ui5/webcomponents-fiori
```

Now import the ui5-button component in the main.js file to be able
to use it (Listing 4-7).

Listing 4-7. Import UI5 Web Components

```
import "./style.css";
import "@ui5/webcomponents/dist/Button"; // ui5-button
```

The page should reload, and, this time, you should see an SAP Fiori–compliant button on the page as depicted in Figure 4-3.

Figure 4-3. *A website with an SAP Fiori button*

Later, we want to make use of icons as well. **So, we need to install these icon packages too (Listing 4-8).**

Listing 4-8. Install icons from UI5 Web Components

```
npm add @ui5/webcomponents-icons @ui5/webcomponents-icons-
business-suite
```

Add More Components

At the time of writing, there were about 100 components available in both modules. The UI5 Web Components Playground[8] lists all of them with the associated API documentation and sample code. For our application, we'll use a subset of them – namely, a shell bar for the header, breadcrumbs to indicate navigation options, a collapsible panel that includes the inputs and submit button of the form, and a list with list items.

Open the '**index.html**' file and replace the entire content of the '**<body>**' tag as follows (Listing 4-9):

Listing 4-9. Add more UI5 Web Components

```
<ui5-shellbar id="shellbar" primary-title="SAP UI Frameworks
for Enterprise Developers: A Practical Guide"
 secondary-title="UI5 Web Components" notifications-count="60"
 show-notifications show-product-switch>
 <ui5-avatar slot="profile" initials="MV"></ui5-avatar>
 <img slot="logo" src="https://sapui5.hana.ondemand.com/
 resources/sap/ui/documentation/sdk/images/logo_ui5.png" />
</ui5-shellbar>
<div id="page">
 <ui5-breadcrumbs>
   <ui5-breadcrumbs-item>Drinks</ui5-breadcrumbs-item>
   <ui5-breadcrumbs-item>Beer</ui5-breadcrumbs-item>
 </ui5-breadcrumbs>
 <ui5-panel width="100%" header-text="Add new beer"
 class="full-width" collapsed>
   <ui5-input id="name" required placeholder="Name" show-clear-
   icon></ui5-input>
```

---

[8] https://sap.github.io/ui5-webcomponents/playground/components

```
  <ui5-input id="abv" placeholder="ABV" type="Number"
  required></ui5-input>
  <ui5-input id="ibu" placeholder="IBU" type="Number"
  required></ui5-input>
  <ui5-button id="addBeer" design="Emphasized">Add Beer</
  ui5-button>
 </ui5-panel>
 <div id="app">
  <ui5-list id="myList" class="full-width" no-data-text="No
  Beers here :("> </ui5-list>
 </div>
</div>
<script type="module" src="/main.js"></script>
```

You probably noticed that the website doesn't look as expected. Similar to before, this happened due to the missing import statements. **You can fix this by adding the snippet in Listing 4-10 at the bottom of the main.js file**.

Listing 4-10. Import more UI5 Web Components at runtime

```
import "@ui5/webcomponents-fiori/dist/ShellBar"; //
ui5-shellbar
import "@ui5/webcomponents/dist/Avatar"; // ui5-avatar

import "@ui5/webcomponents/dist/Link"; // ui5-link
import "@ui5/webcomponents/dist/Breadcrumbs"; // ui5-
breadcrumbs

import "@ui5/webcomponents/dist/Panel"; // ui5-panel
import "@ui5/webcomponents/dist/Button"; // ui5-button
import "@ui5/webcomponents/dist/Input"; // ui5-input

import "@ui5/webcomponents/dist/List"; // ui5-list
import "@ui5/webcomponents/dist/StandardListItem"; // ui5-li
```

```
import "@ui5/webcomponents-icons/dist/AllIcons";
import "@ui5/webcomponents-icons-business-suite/dist/bottle";
// bottle icon
```

Now you can see the SAP Fiori user interface shown in Figure 4-4. You can expand and collapse the "Add new beer" form, and the list is still empty. It's worth noting that while this user interface uses UI5 Web Components, it's not following the best practices defined in the SAP Fiori guidelines. The form fields don't have labels and use the input's placeholder to describe the desired input. This can be seen as a disadvantage *and* an advantage. For some applications, it might not be necessary to be 100% SAP Fiori compliant and rather benefit from the freedom this approach provides to your designers. This was, to some degree, already possible in UI5, but UI5 Web Components are pushing the boundaries here. However, with this freedom comes a responsibility to define your own consistent UX within the application.

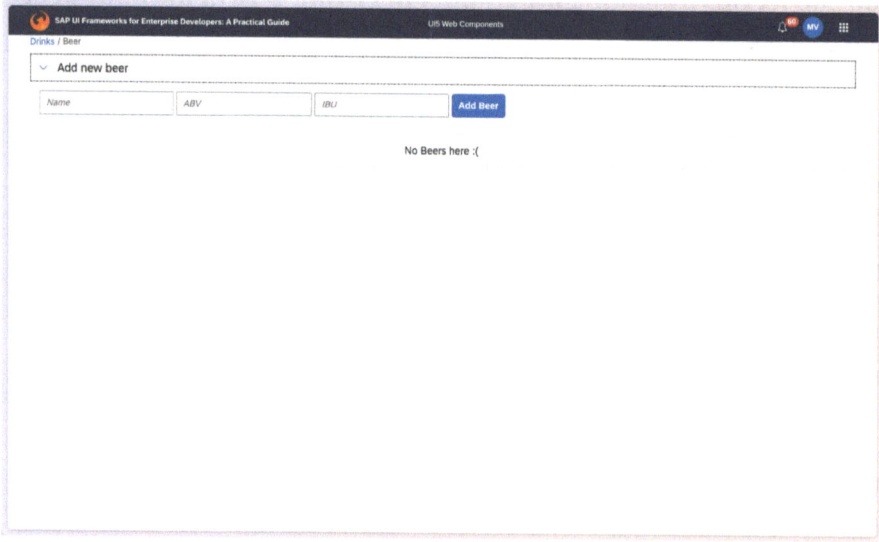

Figure 4-4. *An empty list with the form*

Display Sample Data

Let's bring some life to the list and render list items for the beverages that are returned from an HTTP request. As we're using vanilla JavaScript, there is no concept of data binding here. But as mentioned before, we can implement anything that's missing on our own. This implementation can be as small as a function that gets the DOM element of the list control and an array with all the data entries. The function uses the reduce() method to loop through each item in the array and generate an HTML string for each item. The generated HTML includes a ui5-li element with the beer name, icon, and information about the IBU and ABV levels of the beer.

Finally, the innerHTML property of the parent element is set to the generated HTML string, binding the list of beer items to the parent element.

Create a new file called databinding.js with the content in Listing 4-11.

Listing 4-11. Emulating data binding with UI5 Web Components

```
export function bindBeerItems(parent, list) {
    const items = list.reduce((markup, item) => {
        return (
            markup +
            `<ui5-li icon="business-suite/bottle"
            description="IBU: ${item.ibu} ABV ${item.abv
            }" additional-text="
        ${item.ibu > 25 ? "Very bitter" : item.ibu > 10 ?
        "Bitter" : ""
            }" additional-text-state="${item.ibu > 25 ? "Error"
            : item.ibu > 10 ? "Warning" : ""
            }">${item.name}</ui5-li>`
        );
    }, "");
    parent.innerHTML = items;
};
```

185

We could write this code in any existing or new file, since we don't use a framework. This lack of structure can be a problem for projects that are larger than this short exercise. If you want to use this technology in production, we recommend that you define your own folder structure for several reasons:

1. *Organization and readability*: Separating code files into different folders based on their functionality or purpose makes it easier to find and understand specific code when working on a project. This can save time and reduce errors caused by confusion or disorganization.

2. *Maintainability and scalability*: As a project grows, the number of code files and modules also increases. Defining a clear folder structure from the start helps to keep code organized and maintainable as the project scales. It also allows new team members to quickly understand the project structure and contribute to it.

3. *Ease of collaboration*: A clear and consistent folder structure makes it easier for multiple developers to work on different parts of a project at the same time, without causing conflicts or confusion.

4. *Separation of concerns*: By separating code files into different folders based on their functionality, developers can follow the principle of separation of concerns, which means that each module or component should have a specific responsibility and should not overlap with other modules or components. This makes the code more modular and easier to test, debug, and refactor.

With the data binding function in place, it's time to fetch the sample data from localhost. **Add the lines in Listing 4-12 at the end of the main.js file**.

Listing 4-12. Data retrieval and binding with vanilla JavaScript

```
import { bindBeerItems } from "./databinding";

let beers = [];

fetch("/Beers.json")
  .then((response) => {
    if (response.status === 200) {
      return response.json();
    } else {
      throw new Error(response.statusText);
    }
  })
  .then((data) => {
    beers = data;
    bindBeerItems(document.querySelector("#myList"), beers);
  })
  .catch((err) => console.error(err));
```

We won't consider error handling for now and only log a message to the console to keep this exercise short and concise. But error handling would be helpful to find out when the request is failing – like in our case now. This happens because we don't have a local file with sample data. Let's use the Vite Server Options[9] to proxy certain requests to another domain. To accomplish this, **create a new vite.config.js file and add the configuration in Listing 4-13**.

---

[9] https://vitejs.dev/config/server-options.html#server-proxy

Listing 4-13. Create a local data source

```
export default {
    server: {
        proxy: {
            "/Beers.json": "https://raw.githubusercontent.com/
            apress/SAP-UI-Frameworks-for-Enterprise-Developers-
            by-Marius-Obert-Volker-Buzek/main/apps/fiori-
            elements/webapp/localService/data/"
        }
    }
};
```

Vite won't automatically pick up the configuration file. So, you need to stop the current development server and restart it to proxy requests to /Beers.json with npm run dev.

Now you should see a list with multiple entities as depicted in the screenshot shown in Figure 4-5.

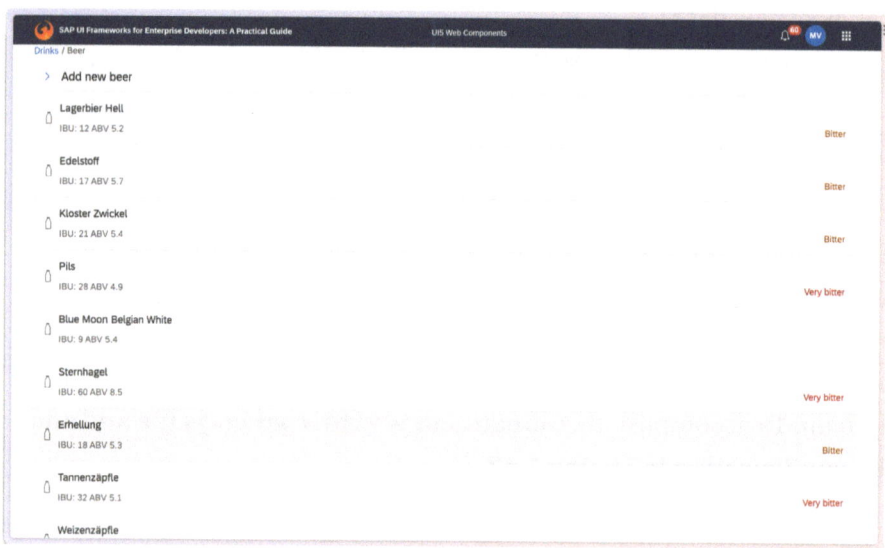

Figure 4-5. The list showing sample data

Handle Form Submissions

The list is already nice, but something is missing: the option to add your own favorite beers! So let's implement the form submission. The first thing we need to do is to register a callback that is triggered when the submit button is clicked. Then, we will capture all the input fields based on their ID and validate the input of these fields. If they are good, we update the beers array, which we'll use as temporary storage, and update the list item binding. Lastly, we'll then reset the form to allow new entries.

If one of the input fields contains empty values, we manually change the value state of this input to Error to highlight it to the user. For further assistance, we also add a value state message.

Append the lines in Listing 4-14 to main.js to implement this logic.

Listing 4-14. Form submission logic

```
document.querySelector("#addBeer").addEventListener("click",
function onSubmit() {
 const name = document.querySelector("#name");
 const ibu = document.querySelector("#ibu");
 const abv = document.querySelector("#abv");

 let allValid = true;

 if (!name.value) {
   name.valueState = "Error";
   name.innerHTML = `<div slot="valueStateMessage">Please add a
   name for this beer.</div>`;
   allValid = false;
 } else {
   name.valueState = "None";
 }
```

```
if (!abv.value) {
  abv.valueState = "Error";
  abv.innerHTML = `<div slot="valueStateMessage">Please add a
  numeric value for the Alcohol By Volume of this beer.</div>`;
  allValid = false;
} else {
  abv.valueState = "None";
}
if (!ibu.value) {
  ibu.valueState = "Error";
  ibu.innerHTML = `<div slot="valueStateMessage">Please add a
  numeric value for the International Bitterness Unit of this
  beer.</div>`;
  allValid = false;
} else {
  ibu.valueState = "None";
}

if (!allValid) {
  return;
}

beers = [{
  name: name.value,
  ibu: ibu.value,
  abv: abv.value
}, ...beers];
bindBeerItems(document.querySelector("#myList"), beers);

name.value = "";
ibu.value = "";
abv.value = "";
});
```

With these changes, you should be able to add new items to the list via the form (Figure 4-6).

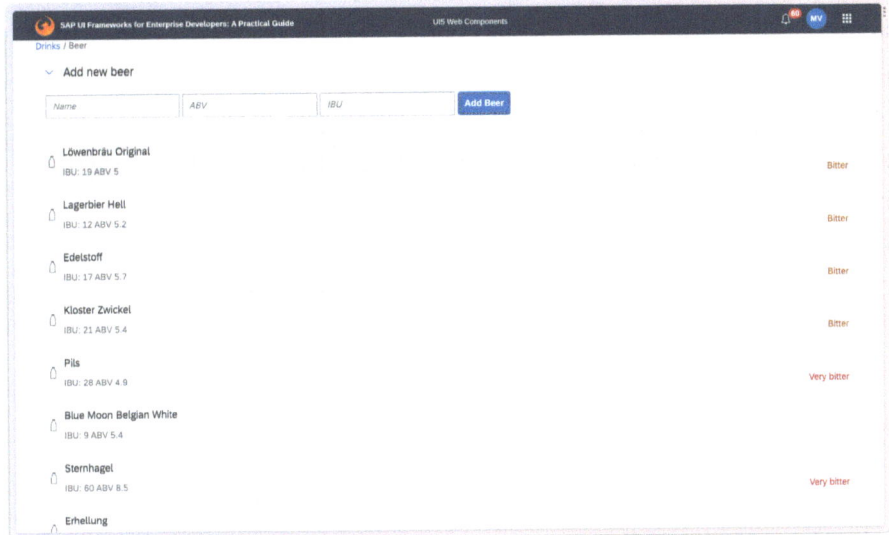

Figure 4-6. *A custom entry (Löwenbräu Original) in the list – all rendered with UI5 Web Components*

This section also illustrated another trait of vanilla JavaScript development. Basic input validation support is provided by the `type` property of the input elements. But advanced input validation and state management need to be handled entirely by ourselves. The same would go for internationalization and formatters if we wanted to use such features too.

Build

Vite already prepared the build script for compiling and optimizing the application's source code for production deployment.

The Vite build script first reads your project's `vite.config.js` file to determine the build configuration. This file can define settings such as the entry point of your application, the output directory, and any plugins or customizations to apply during the build process. It then analyzes your project's source code and generates a dependency graph to determine which files need to be included in the final build. Once Vite has generated the dependency graph, it transpiles your source code from modern JavaScript (ES6+) to older versions of JavaScript that are compatible with a wider range of browsers. The tool also bundles your code into smaller, more efficient files that can be served to the browser. Another benefit of Vite is the built-in optimizations that can be applied during the build process, such as minification, dead code elimination, and tree shaking. These optimizations help to reduce the size of your application's code and improve its performance.

Finally, Vite outputs the optimized and bundled files to the directory specified in your `vite.config.js` file.

As we're using Web Components, we need to tell Vite the component prefix so that it can consider these dependencies in the build process. **Modify the `vite.config.js` as shown in Listing 4-15.**

Listing 4-15. Vite build optimizing UI5 Web Components

```
export default {
    plugins: [
        {
            template: {
                compilerOptions: {
                    isCustomElement: (tag) => tag.
                    startsWith("ui5-")
                }
            }
        }
    ],
    server: {
        proxy: {
```

```
      "/Beers.json": "https://raw.githubusercontent.com/
      apress/SAP-UI-Frameworks-for-Enterprise-Developers-
      by-Marius-Obert-Volker-Buzek/main/apps/fiori-
      elements/webapp/localService/data/"
    }
  }
};
```

All the terminal commands used in this section are for Unix systems like macOS and Linux. If you're using Windows, you'll need to use the Windows Subsystem for Linux to access those same commands. Windows Subsystem for Linux provides a Linux-compatible environment that you can use to run many Unix tools and applications on your Windows machine.

We forwarded all requests to the Beers.json data source to GitHub during development. Once the project has been built, this proxy no longer exists. **Therefore, we need to modify the build script in the package.json to download the data source in the dist folder (Listing 4-16).**

Listing 4-16. Enhanced Vite configuration for data proxying

```
"scripts": {
    ...,
    "build": "vite build && curl https://raw.githubusercontent.
    com/apress/SAP-UI-Frameworks-for-Enterprise-Developers-by-
    Marius-Obert-Volker-Buzek/main/apps/fiori-elements/webapp/
    localService/data/Beers.json > dist/Beers.json",
    ...
  },
```

With this, the build process can be triggered with the command npm run build.

The files in the output folder, dist/ by default, can then be deployed to a web server or a static hosting service to serve your application to users.

A Note on Debugging, Testing, and Deployment

This exercise was intended to show hands-on what UI5 Web Components are and how they can be used. As already mentioned a couple of times throughout this chapter, they can be combined with any web technology tool, framework, and platform. As the developer or architect, you have the freedom **and** responsibility to pick the right tool for the job. Depending on this selection, you also need to pick the optimal tools and processes for debugging, testing, and deployment, a.k.a. CI/CD.

Usage with SPA Frameworks

To make your life easier and boost development efficiency, we recommend using UI5 Web Components with a web framework of your choice. Frameworks often come with an opinionated development pattern and development processes that help you build better solutions. They already suggest best practices on how to deal with error messages, routing, storing ephemeral and persistent data, cover strategies for input validation and state management, and much more. And thanks to its strict adherence to open web standards, UI5 Web Components can be combined with virtually any framework – be it an MVC- or component-based architecture.

For illustration, we mention two frameworks and briefly show how you can embed UI5 Web Components in them.

Usage in UI5

When you think about frameworks you possibly want to combine with UI5 Web Components, you might first think of SAPUI5 and OpenUI5. It makes sense because these two frameworks might already be in use in your company. So far, UI5 controls did a good job abstracting HTML and CSS details from developers and linking the HTML and CSS world with the UI5 programming model. So, where do Web Components fit in as the problem of linking these two worlds is already solved?

This is actually quite similar to the problem of integrating third-party controls, such as Chart.js. This is a requirement that UI5 apps have often faced before, and as a solution, it became a best practice to write custom UI5 controls. These custom controls help to render the markup and properly connect to the rendering life cycle, control life cycle, and data binding of the third-party library. For Web Components, UI5 is doing the same thing. To make the reuse easier, UI5 has been extended with special base classes. These classes take care of the rendering of the custom tag, managing the properties, assigning the aggregations to slots, and registering the custom events. But still, every Web Component which should be used in UI5 needs to be wrapped by a so-called wrapper UI5 control. Fortunately, all of these are already built right into UI5 itself.

The UI5 library that provides the base classes can be found in the namespace `sap.ui.webc.common`. The wrapper controls themselves are part of one out of two additional UI5 libraries:

- `sap.ui.webc.main`[10] provides the bread and butter UI Web Components, a.k.a. `@ui5/webcomponents`.

- `sap.ui.webc.fiori`[11] provides the SAP Fiori–specific Web Components, a.k.a. `@ui5/webcomponents-fiori`.

---

[10] https://openui5.hana.ondemand.com/api/sap.ui.webc.main
[11] https://openui5.hana.ondemand.com/api/sap.ui.webc.fiori

These libraries are optional and can be added to the bootstrap process if needed. But with the availability of these libraries in UI5, some controls are available in two libraries (`sap.m.Button` vs. `sap.ui.webc.main.Button`), which raises the pragmatic question:

Which UI5 control should I use?

The choice of which button control to use depends on the scenario. If you are developing UI5 applications in the context of the SAP Build Work Zone, standard edition, which itself uses `sap.m` UI5 controls, you should continue to use `sap.m` UI5 controls. If you are starting a new UI5 application from scratch for a stand-alone scenario, you can use `sap.ui.webc.main` UI5 controls. For `sap.m` and `sap.ui.webc.main`, it's an either-or decision, but other UI5 libraries can be combined. For example, it makes sense to use `sap.ui.layout`-based layout with UI elements from `sap.ui.webc.main` to create a proper user interface, since UI5 Web Components do not and will not provide many layout components.

SAP plans,[12] going forward, to develop new UI5 controls as UI5 Web Components and then add them back to UI5 via the UI5 Web Components enablement. This makes sense as Web Components are more lightweight (in terms of runtime resource usage) than traditional, JavaScript-based UI5 controls. They help to increase rendering performance of UI5 apps.

Listing 4-17 shows how you can bootstrap an OpenUI5 app, load the additional libraries, and use the UI5 Web Components wrapper – all in **a single file**.

Listing 4-17. A minimal example of using UI5 Web Components with UI5

```
<!DOCTYPE html>
<html>
  <head>
```

---

[12] https://blogs.sap.com/2022/03/10/ui5-web-components-enablement-for-openui5-sapui5/

```
    <meta http-equiv="X-UA-Compatible" content="IE=edge" />
    <meta charset="utf-8" />
    <title>UI5 Web Components Enablement for OpenUI5</title>
    <script
      id="sap-ui-bootstrap"
      src="https://sdk.openui5.org/resources/sap-ui-core.js"
      data-sap-ui-theme="sap_horizon"
      data-sap-ui-libs="sap.ui.webc.main"
      data-sap-ui-async="true"
      data-sap-ui-compatVersion="edge"
    ></script>

    <script>
      sap.ui.getCore().attachInit(function () {
        sap.ui.require(["sap/ui/webc/main/Button"], function
        (Button) {
          new Button({
            text: "Hello World",
          }).placeAt("content");
        });
      });
    </script>
  </head>
  <body id="content" class="sapUiBody"></body>
</html>
```

The code in Listing 4-17 for using the Button from the UI5 Web
Components is no different than using an sap.m.Button. The API in UI5
for using UI5 Web Components looks the same as for traditional UI5
controls. This retrofit alignment needs special considerations by you
as a developer, though. Even though the programmatic usage of both
Web Components and controls is the same (new Button({...})), their

properties might vary slightly. Accustomed attributes are sometimes named differently, though they are surfaced by seemingly equivalent controls.

As an example, the ShellBar from sap.ui.webc.fiori and sap.f looks and feels the same (Figure 4-7).

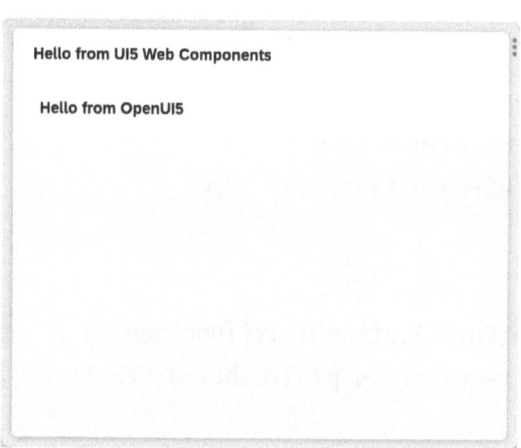

Figure 4-7. *ShellBar from sap.ui.webc.fiori and sap.f*

Yet, notice the difference in the main title attribute: primaryTitle (Web Component) vs. title (control) in Listing 4-18.

Listing 4-18. Property name differences

```
<!DOCTYPE html>
<html>
<head>
    <meta http-equiv='X-UA-Compatible' content='IE=edge'>
    <meta charset="utf-8">
    <title>UI5 Web Components Enablement for OpenUI5</title>
    <script id='sap-ui-bootstrap'
            src='https://sdk.openui5.org/resources/sap-
            ui-core.js'
```

```
        data-sap-ui-theme='sap_horizon'
        data-sap-ui-libs='sap.ui.webc.main,sap.f'
        data-sap-ui-async='true'
        data-sap-ui-compatVersion='edge'></script>

    <script>
        sap.ui.getCore().attachInit(function () {
            sap.ui.require(["sap/ui/webc/fiori/ShellBar",
            "sap/f/ShellBar"], function (wcShellBar,
            fShellBar) {
                new wcShellBar({
                    primaryTitle: "Hello from UI5 Web
                    Components",
                }).placeAt("content")

                new fShellBar({
                    title: "Hello from OpenUI5",
                }).placeAt("content")
            })
        });
    </script>
</head>

<body id='content' class='sapUiBody'></body>
</html>
```

So just be aware that UI5 Web Components and traditional UI5 controls are often no drop-in replacement for the other and require your attention!

Usage in React

Similar to UI5, when it comes to integrating Web Components with React, there are some challenges that need to be addressed. React has its own way of rendering components, and it doesn't natively support Web Components. This means that Web Components cannot be used directly in a React application without some additional work.

This is where the React wrapper for Web Components comes into play. The wrapper is essentially a bridge that connects the Web Component to the React application. It is itself a React component that provides a way to render the Web Component within the React component tree, and it also handles communication between the Web Component and the React application.

"UI5 Web Components for React" is a UI component library that provides a collection of prebuilt, reusable user interface React components. These components are built using the UI5 Web Components library we already know. Since each component needs to be wrapped first, there might be a difference between the components available in @ui5/webcomponents and @ui5/webcomponents-react. We recommend consulting the UI5 Web Components for React Storybook[13] to find a complete list of all available components.

We again use Vite to initialize a new project with the name ui5-webcomponents-react and using the React framework and JavaScript variant. Then install the required dependencies, including the @ui5/webcomponents-react package, which provides the Web Components wrappers. Finally, start the development server for the project, allowing for real-time updates during development (Listing 4-19).

Listing 4-19. Start a project with UI5 Web Components with React

```
npm init vite // name=ui5-webcomponents-react framework=React
variant=JavaScript
cd ui5-webcomponents-react
npm install
npm add @ui5/webcomponents-react
npm run dev
```

---

[13] https://sap.github.io/ui5-webcomponents-react

The default application comes with a few CSS classes that we don't need. Go to src/index.css and override the file with the content in Listing 4-20 that makes sure we use the full width and height we need later.

Listing 4-20. Minimal CSS for the React app

```
body {
  margin: 0;
  display: flex;
  min-width: 320px;
  min-height: 100vh;
}

body > div {
  width: 100%;
}
```

The original sample application uses React states to keep track of a counter that increases when the button is clicked. Let's do the same with UI5 Web Components instead of native HTML elements.

Listing 4-21 shows the content of src/App.jsx which renders a UI5 ShellBar component, along with a Text component showing the current count and a Button component that increments the count when clicked. The component will, same as the generated code, import and use the useState hook from React to manage state, allowing for the count to be updated dynamically.

Listing 4-21. Reactive behavior with a UI5 Web Component

```
import { useState } from "react";
import {
 ThemeProvider,
 ShellBar,
 Button,
```

```
Text,
Avatar,
} from "@ui5/webcomponents-react";

function App() {
 const [count, setCount] = useState(0);

 return (
   <ThemeProvider>
     <ShellBar
       logo={
         <img
           alt="UI5 Logo"
           src="https://sdk.openui5.org/resources/sap/ui/
           documentation/sdk/images/logo_ui5.png"
         />
       }
       primaryTitle="SAP UI Frameworks for Enterprise
       Developers: A Practical Guide"
       secondaryTitle="UI5 Web Components for React"
       showNotifications
       notificationsCount="60"
       profile={<Avatar initials="MV" />}
       showProductSwitch
     />
     <Text>Current count: {count}</Text>
     <Button
       icon="add"
       onClick={function () {
         setCount(count + 1);
       }}
     >
```

```
        Click to increase the counter
      </Button>
    </ThemeProvider>
  );
}

export default App;
```

The preview in your browser should automatically update and show you the result you also see in Figure 4-8.

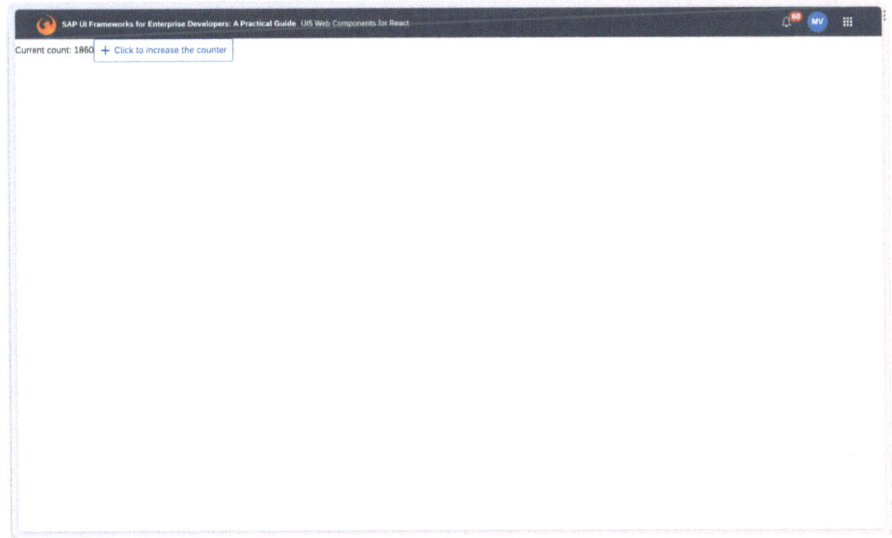

Figure 4-8. *The counter app, built with UI5 Web Components for React*

The Road from Here

The previous exercises showed how to use UI5 Web Components and which features they provide. They are basically universally embeddable in modern web applications, either with or without a wrapper. This opens up a large spectrum of technologies that were previously not usable for enterprise developers. At the same time, this freedom comes with new responsibilities of picking and combining the best-of-breed frameworks and technologies into a successful technology stack.

If this sounds interesting to you, and you would like to explore this technology in more detail, we recommend that you check out the full catalog of available UI5 Web Components.[14] Make sure to bookmark this page and regularly check which new components have been added. Similarly, if you want to use a wrapper implementation for an SPA framework, such as UI5,[15] React,[16] or Angular,[17] make sure there are wrappers for the controls you intend to use. If the component you are looking for does not exist, consider implementing a custom UI5 Web Components package.[18]

Once you are familiar with UI5 Web Components, it's time to explore tech stack candidates you might want to consider. The most famous SPA frameworks out there are currently React,[19] Angular,[20] Vue,[21] and Svelte[22] – just to name a few. The number of testing frameworks, build tools, and deployment targets goes on, and the number of possible combinations is almost infinite. So, it's best not to rush this decision and instead prototype with all technologies for a while until you make this call. Depending on the size of the project you are working on, it makes sense to limit the number of vendors/maintainers of the technologies to avoid becoming

---

[14] https://sap.github.io/ui5-webcomponents/playground/components

[15] https://openui5.hana.ondemand.com/api/sap.ui.webc.main

[16] https://sap.github.io/ui5-webcomponents-react/?path=/docs/getting-started--docs

[17] https://github.com/SAP/ui5-webcomponents-ngx

[18] https://sap.github.io/ui5-webcomponents/playground/development/custom-ui5-web-components-packages/

[19] https://reactjs.org/

[20] https://angular.io/

[21] https://vuejs.org/

[22] https://svelte.dev/

dependent on a deprecated open source project. Using too many different technologies usually also leads to a higher total cost of ownership and could pave the way for technical debt.

Similarly to the layers of your technology stack, it is worth investing time in finding the right IDE extensions, for example, from the VS Code Marketplace,[23] that work with your development processes and pipeline to efficiently build outstanding products.

UI5 Web Components for Enterprise Applications

You have seen the advantages that UI5 Web Components can bring to enterprise software. They are built with the latest web technologies and leverage modern browser features without having to worry about the weight of technical debt and legacy dependencies. Still, they implement the well-known SAP Fiori design system and make the behavior of the most popular controls available to any SPA framework out there. And all of this is done with open source and a transparent development process to make sure this technology is future-proof.

The individual components come with the accessibility, globalization, and security aspects they need. However, this still means that it's up to the developer to make sure the full application is also accessible, secure, and integrates with all backends it has to.

Additionally, the developers and architects have more responsibilities to handle. Most of the requirements we discussed in Chapter 1 (like scalability, maintainability, and compatibility) don't come with UI5 Web Components themselves. In fact, maintainability can become a bit harder because you don't depend on one vendor, SAP, but a number of software vendors, some

---

[23] https://marketplace.visualstudio.com/vscode

of which are open source developers. It's up to you to pick and integrate libraries and tools into your tech stack or to implement the tooling yourself.

This is in many ways the opposite of what SAP Fiori elements and UI5 offer and, therefore, might not automatically be the right choice for your next project – especially if it's a "traditional" enterprise application project. However, this is not just a chance to upgrade and modernize your technology stack, it is also an opportunity to rethink your current development process. This technology gives you the chance to hire frontend engineers with a different skill set and also provides enough freedom to hire UX engineers and designers if you don't have these roles in your engineering team as of now. At the time of this writing, there are not many established best practices available, which is also a good opportunity to join the larger community of SAP UI developers and share your knowledge. If you are not confident how to do that, jump to Chapter 7 to learn more.

Another shortcoming of UI5 Web Components lies in the nature of Web Components themselves. They are reusable, but have their entire inner workings shielded away in a black box, the shadow DOM. With this, they provide a "take it or leave it" offer. There is no way to modify or extend their inner mechanics that goes beyond the predefined interfaces, such as CSS variables, CSS shadow parts, slots, or extension points. Further, you can only add aggregations within Web Components via so-called predefined slots.[24] As a consequence, it might be hard or even impossible to define Web Components for certain kinds of layouts or forms for which you don't exactly know their inner structures. But luckily, there is another, more fine-granular technology option that provides this freedom and doesn't hide its internal implementation in the shadow DOM. And the best of it all is that you can mix and match between this technology and UI5 Web Components. We'll explore this option in the next chapter.

---

[24] https://developer.mozilla.org/en-US/docs/Web/Web_Components/
Using_templates_and_slots

CHAPTER 5

Elementary: Fundamental Library Styles

In this chapter, we will show you how multiple frameworks bring famous design systems to the presentation layer of the Web: CSS. We'll begin with an inspection of current CSS frameworks and the benefits they offer to web developers.

Then, we'll guide you on seamlessly integrating the open source project Fundamental Library Styles into an existing web application, ensuring a smooth and cohesive SAP Fiori user experience. You'll discover how to leverage the predefined styles provided by the library to enhance the visual appeal and consistency of your web designs.

We'll conclude by reflecting on the potential of Fundamental Library Styles for enterprise applications and discussing how they can promote consistency across legacy and other web applications.

© Marius Obert and Volker Buzek 2023   207
M. Obert and V. Buzek, *SAP UI Frameworks for Enterprise Developers*,
https://doi.org/10.1007/978-1-4842-9535-9_5

What Are Fundamental Library Styles?

Fundamental Library Styles[1] have a few things in common with UI5 Web Components. They are also open source, don't offer a programming model, and offer the developers to bring their favorite web development framework. Another parallel is they also provide a set of UI components such as buttons, date pickers, and file uploaders. But while the UI5 Web Components came with the associated behavior logic of these components, **Fundamental Library Styles limits itself to the visual appearance**. You might also call this appearance the *fundamental styles*. These styles come in the shape of Cascading Style Sheet (CSS) classes only – without any JavaScript. As a result, the components are missing any behavior that you might expect. For example, the date picker doesn't have the functionality to expand, collapse, or change the displayed data, and the drop-down menu cannot show its options without any additional code from the developer.

This might sound confusing, and you probably wonder *why* someone wants such a technology. For once, this makes the technology even more lightweight than Web Components, and it offers maximum flexibility to developers and designers. There are a couple of other very popular CSS frameworks out there that follow the same approach. Bootstrap,[2] Materialize CSS,[3] and Tailwind CSS[4] are just a few popular examples of this category that get tens of millions of downloads every week. One big advantage is that CSS libraries shield developers from writing their own class definitions and worrying about consistency and cross-browser compatibility. At the same time, they allow developers to use them as a starting point and build with any web technology on top of it.

---

[1] https://sap.github.io/fundamental-styles/
[2] https://getbootstrap.com/docs/5.3/components/accordion/
[3] https://materializecss.com/
[4] https://tailwindcss.com/

While Web Components are, by design, not very customizable, Fundamental Library Styles is highly customizable and more open. This means that the technology can also provide layouting support, which intentionally wasn't possible with UI5 Web Components. One of the largest benefits of Fundamental Library Styles is that it's so lightweight that you can even add it on top of an existing web app without adding too many new dependencies. Because of this, Fundamental Library Styles are often used to add an "SAP Fiori layer" on top of existing web applications. Other use cases are proof of concepts (POCs) and prototypes when designers require more freedom (Figure 5-1). With Fundamental Library Styles, they can work creatively without having to strictly follow the SAP Fiori design systems while, at the same time, making sure the resulting web apps and prototypes still look familiar. It's also worth noting that the General Availability release of Fundamental Library Styles is still pending.

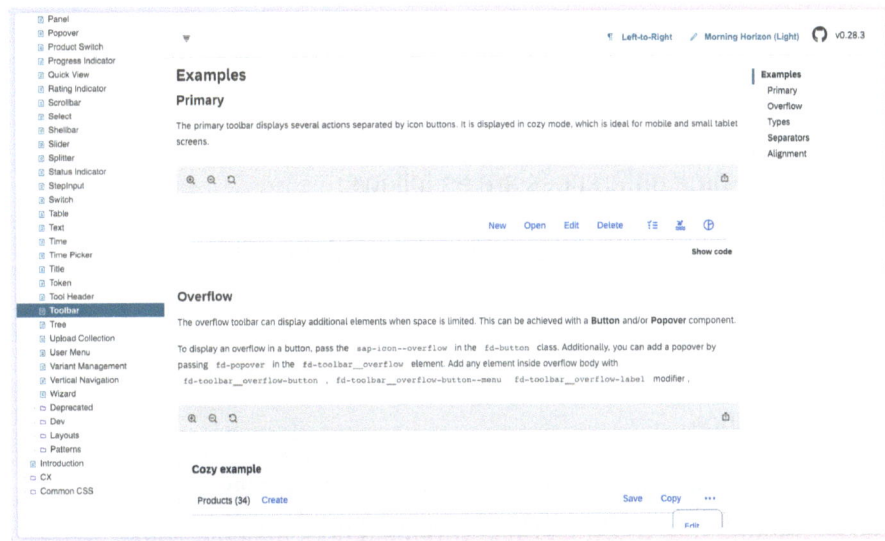

Figure 5-1. *The toolbar styles in action in the Storybook[5]*

---

[5] https://sap.github.io/fundamental-styles/?path=/docs/introduction--styles

Extending CSS

As Fundamental Library Styles are delivered in CSS, it makes sense to recap some advanced techniques that are used to simplify dealing with style sheets, namely, Sass (Syntactically Awesome Style Sheets[6]) and mixins.

Sass is a preprocessor language that extends the capabilities of CSS by providing additional features and functionality. Sass is similar to CSS in terms of syntax, but uses the file ending .scss. It allows you to use features that are not available in standard CSS, such as variables, mixins, nesting, and functions. These features make writing and maintaining CSS code much easier and more efficient.

One of the primary features of Sass is the ability to use variables. In CSS, you often need to use the same value for multiple properties, such as a color or font size. With Sass, you can define a variable once and then use it throughout your code. This makes it easy to update a value across your entire codebase by simply updating the variable.

CSS has variables of its own, which are different from Sass variables! Some differences are as follows:

- Sass variables are eliminated during the Sass compilation process, while CSS variables are retained in the resulting CSS code. So CSS variables can be used to update the styles in all places of a website at once.

- CSS variables allow for distinct values across multiple elements, whereas Sass variables hold a single value at any given time.

---

[6] https://sass-lang.com/

Sass also allows you to use mixins, which are reusable blocks of code that contain a set of CSS properties and values. Mixins can be used to simplify complex CSS code or to provide a consistent style across your entire codebase. Mixins can also take arguments, which allows you to create dynamic styles based on input values.

Nesting is another useful feature of Sass. It allows you to nest CSS selectors inside one another, making it easier to organize your code and avoid repetition. This is especially useful when writing complex styles for large websites or applications.

Sass also provides a range of built-in functions that can be used to perform calculations, manipulate colors, and generate random numbers, among other things. These functions can be extremely useful when working with complex styles or when creating dynamic styles based on user input.

To use Sass, you need to compile your Sass code into CSS. Once your Sass code has been compiled into CSS, you can include it in your HTML code just like regular CSS.

Mixins are reusable blocks of code that can be included in other style sheets. They allow you to define a set of CSS properties and values as a single rule and then reuse that rule in multiple places throughout your codebase.

In essence, mixins are a way of creating your own custom CSS functions. They allow you to define a set of properties and values that can be applied to any selector in your style sheet.

The syntax for defining a mixin in CSS is straightforward. You start by defining a rule with a name, which is typically prefixed with a hyphen (-). This is followed by a set of properties and values that you want to include in your mixin. For example, see Listing 5-1.

Listing 5-1. Defining a Sass mixin

```
@mixin my-mixin {
  font-size: 16px;
  color: #333;
}
```

211

Once you've defined your mixin, you can include it in any other rule using the @include directive, followed by the name of your mixin. For example, see Listing 5-2.

Listing 5-2. Calling a Sass mixin

```
.my-class {
  @include my-mixin;
}
```

This will apply the properties and values defined in your mixin to the my-class selector.

Mixins are especially useful for creating reusable styles for common UI elements, such as buttons or form inputs. By defining a set of properties and values as a mixin, you can easily apply those styles to any element in your codebase, simply by including the mixin in the relevant rule.

Setting Up Your Workspace

If you have completed any of the exercises in the previous chapters, you already have VS Code[7] and Node.js[8] installed. In case you started with this chapter, please install these tools now. Besides them, you should know that Vite[9] is the build tool that we're using for this application. We recommend reading the previous chapter if you would like to learn more about this tool.

---

[7] https://code.visualstudio.com/download

[8] https://nodejs.org/en/download/

[9] https://vitejs.dev/

Learning by Doing

As mentioned before, this technology shows its strength when added on top of other web applications. In that spirit, we'll use this exercise to extend the result of the last exercise. We will take the application that displays the list of beverages and remove the controls that are clearly not in sync with the SAP Fiori design system: the panel that hides the entry form and the breadcrumbs that cannot be used for navigation. Instead, we want to wrap the list in a page that includes a button to navigate back, as well as a header and a footer. The footer will show a button that can be used to clear the list, and the header shows a button that will toggle a popup that includes the form to add new beverages. Overall, this new page will provide a user interface that is more intuitive and familiar to users familiar with the SAP user experience without strictly following the rules of the design system.

Copy the Existing Project

Run the command in Listing 5-3 to copy the contents of the previous exercise in a new directory or clone it from GitHub if you want to start from our template.

All the terminal commands used in this section are for Unix systems like macOS and Linux. If you're using Windows, you'll need to use the Windows Subsystem for Linux to access those same commands. Windows Subsystem for Linux provides a Linux-compatible environment that you can use to run many Unix tools and applications on your Windows machine.

Listing 5-3. Terminal commands to start this project

```
# copy the existing project
cp -r ui5-webcomponents fundamental-library-styles

# or clone the template from the GH
git clone https://github.com/Apress/SAP-UI-Frameworks-for-
Enterprise-Developers-by-Marius-Obert-Volker-Buzek cloned
cp -r cloned/apps/ui5-webcomponents fundamental-library-styles
```

Alternatively, you could also create a new project with Vite or any other web development tooling. But note that the import instructions usually depend on the web development tooling, and your mileage may vary.

Change the **name** field in the **package.json** to **fundamental-library-styles**. Similarly, go to the **index.html** file and adjust the **secondary-title** of the **ui5-shellbar** to **Fundamental Library Styles**.

Once you've done that, navigate to the new folder and **start the development server** to make sure the application is running as expected (Listing 5-4).

Listing 5-4. Running the Fundamental Library Styles project

```
cd fundamental-library-styles
npm install
npm run dev
```

Figure 5-2 shows the result of the previous exercise, but with an updated shell bar.

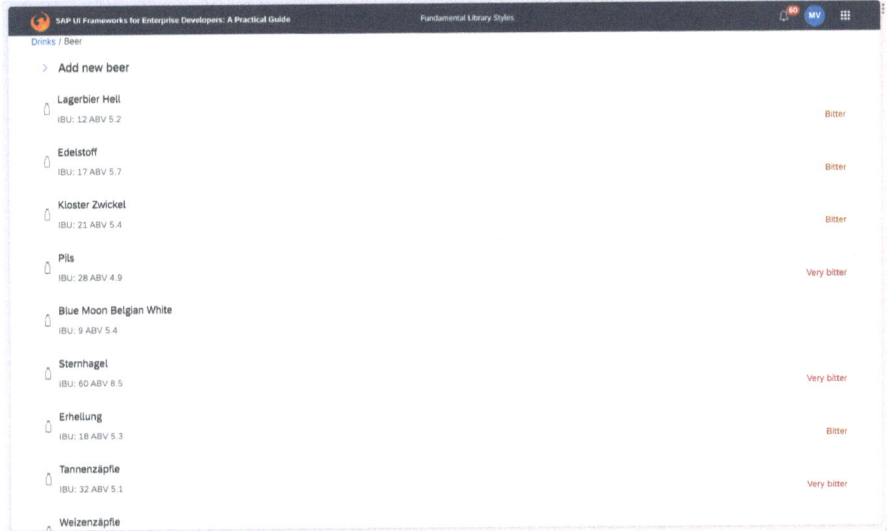

Figure 5-2. *The starting point for this exercise*

Add HTML Elements with Fundamental Styles

The first thing we want to change is the page layout. Currently, we show breadcrumbs that indicate navigation links, and then we show the panel that hides the form and the list. Let's simplify this and **replace the #page div with this new main element (Listing 5-5)**.

Listing 5-5. Simplify the panel and breadcrumb

```
<main class="fd-page">
 <header>
   <div class="fd-bar fd-bar--page fd-bar--header">
     <div class="fd-bar__left">
       <div class="fd-bar__element">
         <button
           class="fd-button fd-button--transparent"
           aria-label="Go back"
         >
```

```
                <i class="sap-icon--navigation-left-arrow"></i>
            </button>
          </div>
          <div class="fd-bar__element">All Beers</div>
        </div>
      </div>
    </header>
    <div class="fd-page__content flex-auto" role="region">
      <div>
        <ui5-list id="myList" no-data-text="No Beers Here :( ">
        </ui5-list>
      </div>
    </div>
    <footer>
      <div class="fd-bar fd-bar--page fd-bar--footer">
        <div class="fd-bar__right">
          <div class="fd-bar__element">
            <button id="clearList" class="fd-button fd-button--
            transparent">
              Clear list
            </button>
          </div>
        </div>
      </div>
    </footer>
</main>
```

Figure 5-3 shows how the updated (and broken) preview page looks.

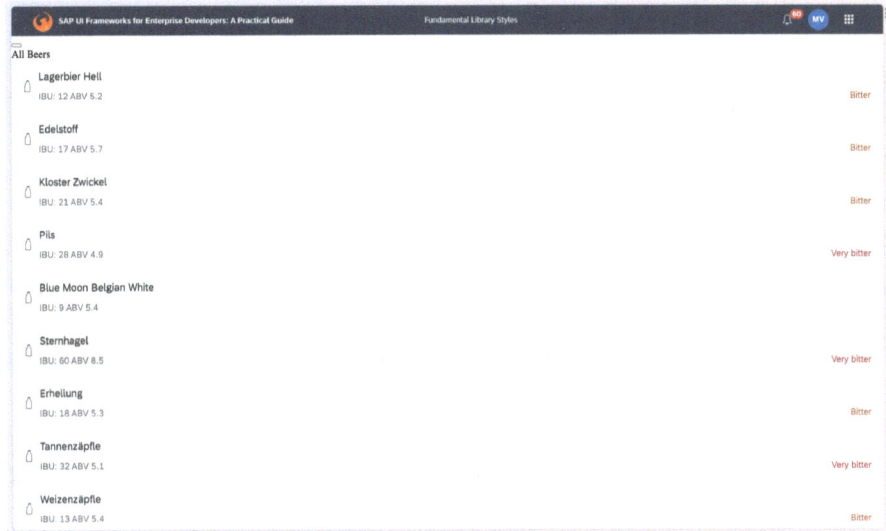

Figure 5-3. *Preview that shows an incomplete website*

Notice that, first, our `main.js` script is running in an error when trying to attach the *form submitted* event to the button that no longer exists (we'll fix that one later). And second, because our project is not yet aware of Fundamental Library Styles. Let's change that!

Integrate Fundamental Library Styles in Your Project

The first step to add Fundamental Library Styles to *any* project is to install it with a package manager. **Run the command in Listing 5-6 in a new terminal session to do this with npm.**

Listing 5-6. Add basic Fundamental modules

```
npm add fundamental-styles @fundamental-styles/common-css
```

Similar to what we had to do in the previous chapter, it's now time to import the style sheets. **Go to the `main.js` file and add these imports under the other import directives (Listing 5-7).**

217

Listing 5-7. Import styles

```
import "fundamental-styles/dist/button.css";
import "fundamental-styles/dist/bar.css";
import "fundamental-styles/dist/page.css";
import "fundamental-styles/dist/icon.css";
import "fundamental-styles/dist/popover.css";
import "fundamental-styles/dist/message-strip.css";
```

This should fix the most urgent styling issues of the page, which should now look similar to what is shown in Figure 5-4.

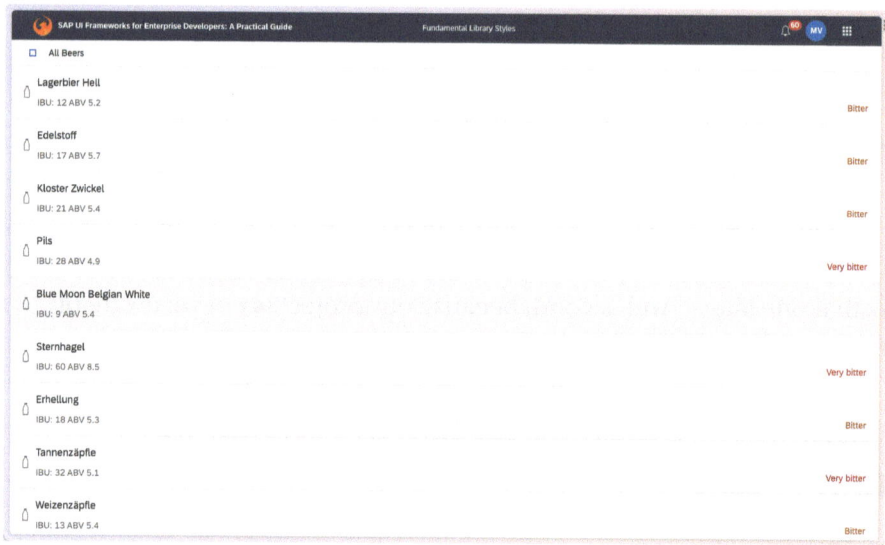

Figure 5-4. *Leveraging UI5 Web Components and Fundamental Library Styles*

But there is still one problem. The icon on the "navigate back" button, in the top-left corner, doesn't render correctly. This happened because we didn't import the web fonts that are used for characters *and* icons. To load them as well, specify custom fonts with which we want to display text. **Append the lines in Listing 5-8 at the end of the style.css**.

Listing 5-8. Enhance the style sheet with the SAP Fiori font 72

```
@font-face {
 font-family: "72";
 src: url("@sap-theming/theming-base-content/content/Base/
 baseLib/baseTheme/fonts/72-Regular-full.woff")
   format("woff");
 font-weight: normal;
 font-style: normal;
}

@font-face {
 font-family: "72";
 src: url("@sap-theming/theming-base-content/content/Base/
 baseLib/baseTheme/fonts/72-Bold-full.woff")
   format("woff");
 font-weight: 700;
 font-style: normal;
}

@font-face {
 font-family: "72";
 src: url("@sap-theming/theming-base-content/content/Base/
 baseLib/baseTheme/fonts/72-Light-full.woff")
   format("woff");
 font-weight: 300;
 font-style: normal;
}

@font-face {
 font-family: "SAP-icons";
 src: url("@sap-theming/theming-base-content/content/Base/
 baseLib/baseTheme/fonts/SAP-icons.woff")
```

```
   format("woff");
 font-weight: normal;
 font-style: normal;
}

@font-face {
 font-family: "BusinessSuiteInAppSymbols";
 src: url("@sap-theming/theming-base-content/content/Base/
 baseLib/baseTheme/fonts/BusinessSuiteInAppSymbols.woff")
    format("woff");
 font-weight: normal;
 font-style: normal;
}

@font-face {
 font-family: "SAP-icons-TNT";
 src: url("@sap-theming/theming-base-content/content/Base/
 baseLib/baseTheme/fonts/SAP-icons-TNT.woff")
    format("woff");
 font-weight: normal;
 font-style: normal;
}
```

Add the dependency to install the fonts (Listing 5-9).

Listing 5-9. Install SAP Fiori dependency to obtain font 72

```
npm add @sap-theming/theming-base-content
```

The page already looks ready because most of the needed CSS variables[10] are already defined by the UI5 Web Components – which comes by default with a theme. However, this is not a guarantee that all variables

---

[10] https://developer.mozilla.org/en-US/docs/Web/CSS/Using_CSS_custom_properties

are initialized as Fundamental Library Styles requires variables that are not defined by UI5 Web Components (and vice versa). Therefore, it's necessary to import the style of the theme manually when working with Fundamental Library Styles. If this was a stand-alone Fundamental Library Styles project, without UI5 Web Components, you would immediately see that something is wrong. **Add the two import statements in Listing 5-10 to main.js** to address this issue and use the SAP Horizon theme.

Listing 5-10. Use SAP Fiori CSS with Fundamental Library Styles

```
import "fundamental-styles/dist/theming/sap_horizon.css";
import "@sap-theming/theming-base-content/content/Base/baseLib/
sap_horizon/css_variables.css";
```

Now the page, also depicted in Figure 5-5, looks just like any other website that uses the SAP Fiori design system. Or can you tell which elements of the page are rendered with which UI technology just by looking at it? We bet you can't because there is no visual difference!

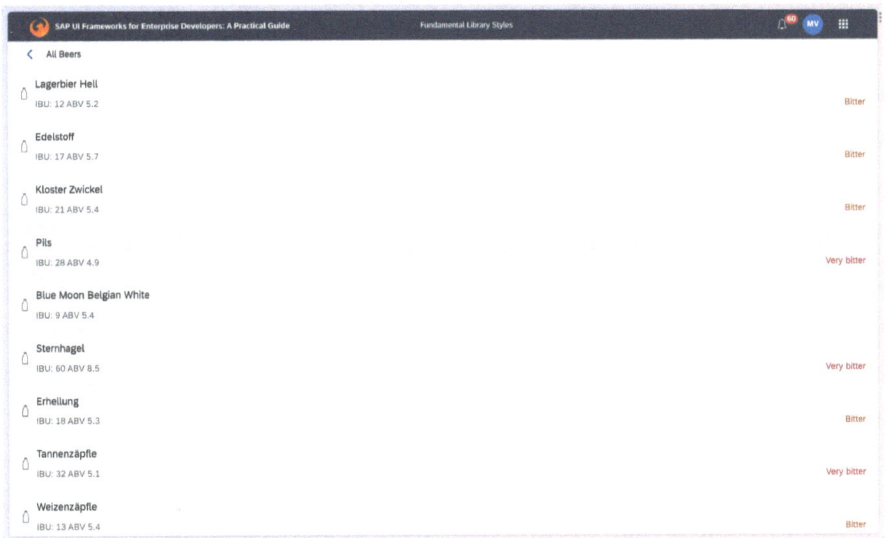

Figure 5-5. *UI5 Web Components and Fundamental Library Styles next to each other in harmony*

221

Clear the List

You have probably noticed there is a "Clear List" button on the bottom of the page that doesn't work yet. Let's change that. As this is a normal HTML button (not a Web Component), we have multiple options to add a callback. To keep things consistent, we attach the callback in our main.js file, in the same fashion that we've added event handlers in the previous chapter. **Add the four lines** in Listing 5-11 to reset the beers array **above** the line in which we'll add the form submission (that is currently broken). It won't work if you add these lines below. The missing form submission button breaks the JavaScript execution and that would mean your "clear list" callback would never get attached (Listing 5-11).

Listing 5-11. Add an event handler for clearing the list

```
document.querySelector("#clearList").
addEventListener("click", () => {
  beers = [];
  bindBeerItems(document.querySelector("#myList"), beers);
});
document.querySelector("#addBeer").addEventListener("click",
function onSubmit() {
...
```

Try the "Clear List" button now to clear all entries. Your list should now be empty and display the "no data" text instead of list elements, as shown in Figure 5-6. Note that a refresh will fetch the list again and, therefore, display all items again. This is expected for this demo. Next, we're going to add the form back to be able to populate the list once it has been cleared. This will also fix the error message we see in the browser console.

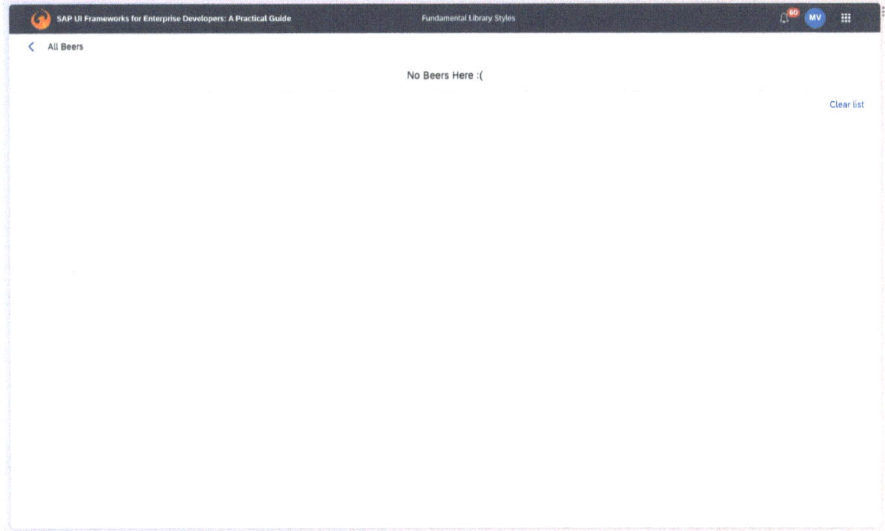

Figure 5-6. *The cleared list with no items*

Bring the Form Back

It's time to address the outstanding issue with the broken main script file. This issue is caused by the missing form that we temporarily removed. Let's bring it back within a popover component that is controlled by a button in the list header. You can make all of these changes, then you **replace the ‹header› element with the code in Listing 5-12.**

Listing 5-12. The add form within the popover

```
<header>
  <div class="fd-bar fd-bar--page fd-bar--header">
    <div class="fd-bar__left">
      <div class="fd-bar__element">
        <button
          class="fd-button fd-button--transparent"
          aria-label="Go back">
          <i class="sap-icon--navigation-left-arrow"></i>
```

```
      </button>
    </div>
    <div class="fd-bar__element">All Beers</div>
  </div>
  <div class="fd-bar__right">
    <div class="fd-bar__element">
      <div class="fd-popover">
        <div class="fd-popover__control">
          <button
    id="addButton"
    aria-controls="popoverForm"
    class="fd-button fd-button--transparent"
    aria-label="Add Beer" >
    <i class="sap-icon--add"></i>
          </button>
        </div>
        <div
          class="fd-popover__body fd-popover__body--right fd-
          popover__body--arrow-x-end"
          aria-hidden="true"
          id="popoverForm" >
          <div class="fd-popover__wrapper">
            <ui5-input id="name" required placeholder="Name"
            show-clear-icon></ui5-input>
            <ui5-input id="abv" required placeholder="ABV"
            type="Number"></ui5-input>
            <ui5-input id="ibu" required placeholder="IBU"
            type="Number"></ui5-input>
            <ui5-button id="addBeer" design="Emphasized">Add
            Beer</ui5-button>
          </div>
        </div>
```

```
      </div>
    </div>
  </div>
</div>
</header>
```

If you have a closer look at line 27, you'll see that property `aria-hidden` is set to `false`. This means the popover is hidden by default. **Add the code in Listing 5-13 at the end of the main.js file** to attach an event listener that toggles the popover every time the add button is clicked.

Listing 5-13. The event handler to show the popover

```
document.querySelector("#addButton").
addEventListener("click", () => {
  const popover = document.querySelector("#popoverForm");
  popover.ariaHidden = popover.ariaHidden === "true" ? "false"
: "true";
});
```

This also means that the only way to toggle the visibility of the popover is via the add button. The SAP Fiori design system says that popovers shall close when a click outside of it is registered, and the popover of UI5 Web Component will always do so without any option for us to ignore that. With Fundamental Library Styles, it's up to us developers to decide when the popover shall be visible, and we have the freedom to diverge from the design system.

The logic to save the new list item is still in place. So let's focus on what's happening when a new beverage has been added: a floating success message in the shape of a message strip[11] component. **Add this markup to the bottom of the index.html file right after the closing </main> tag (Listing 5-14).**

---

[11] https://sap.github.io/fundamental-styles/?path=/docs/components-message-strip--default-strip#success

Listing 5-14. An SAP Fiori–compliant message strip
denoting success

```
</main>
<div
 id="success-strip"
 class="exercise-hidden exercise-float-bottom fd-message-strip
 fd-message-strip--success fd-message-strip--dismissible"
 role="alert"
>
  <p class="fd-message-strip__text">
    The new beer has been added successfully to the list.
  </p>
  <button
    id="closeButton"
    class="fd-button fd-button--transparent fd-button--compact
    fd-message-strip__close"
    aria-controls="success-strip"
    aria-label="Close"
  >
    <i class="sap-icon--decline"></i>
  </button>
</div>
```

The markup snippet uses almost exclusively classes delivered by
Fundamental Library Styles, with two exceptions: `exercise-hidden`
and `exercise-float-bottom`. We'll later use these classes to control the
visibility of the message strip that does not provide a built-in flag, like the
popover did with `aria-hidden`. Other than that, this markup doesn't hold
any surprises and defines a success strip with a text and a close button that
can be used to discard the message strip. **Let's append the event listener
that will remove the custom CSS class and change its visibility to
main.js (Listing 5-15).**

Listing 5-15. Vanilla JavaScript implementation to display the message strip

```
document.querySelector("#closeButton").
addEventListener("click", () => {
 const strip = document.querySelector("#success-strip");
 strip.classList.add("exercise-hidden");
});
```

There's one thing missing. As of now, we never remove the class exercise-hidden which shows the message strip in the first place. Furthermore, it makes sense to define a second condition on which the popover closes: when a new beverage has been added successfully. **Add the code in Listing 5-16 to the end of the onSubmit function** to change the visibility of these two components.

Listing 5-16. Sync popover display with message strip

```
const strip = document.querySelector("#success-strip");
strip.classList.remove("exercise-hidden");

const popover = document.querySelector("#popoverForm");
popover.ariaHidden = popover.ariaHidden === "true" ? "false"
: "true";
```

Now test our latest changes. Click the add button in the page header to show the popover and compare the website you see with Figure 5-7. You should see the same.

227

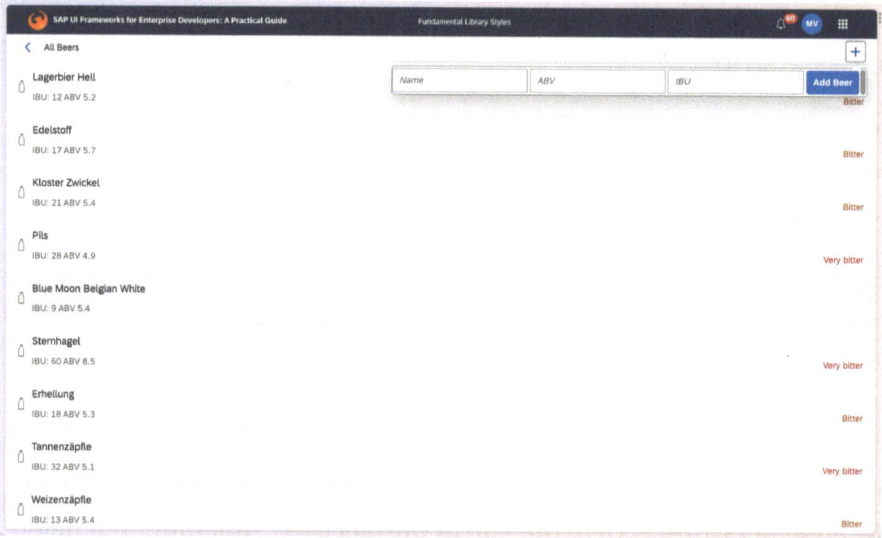

Figure 5-7. *An active popover with a form*

Now complete the form with a new beverage and hit the submit button. The beverage should appear at the top and the message strip below the list (scrolling might be necessary to see it). Don't worry about the message strip that is always visible. We'll take care of that in the next section.

Use Sass Mixins

So far, we only focused on the component-related features of Fundamental Library Styles. But this is not the only aspect that improves the lives of SAP UI developers. There is another package called @fundamental-styles/ common-css that is also part of the project. You might recall that you installed it earlier without knowing its purpose. It provides CSS rules based on SAP Fiori Design Guidelines, which, by the way, can also be used to bring layouting to UI5 Web Components apps. Further customization is possible by using the Sassy Cascading Style Sheets (SCSS) mixins and providing user-defined values as parameters. Figure 5-8 shows the documentation page with the expanded Common CSS tree that exposes all available mixins.

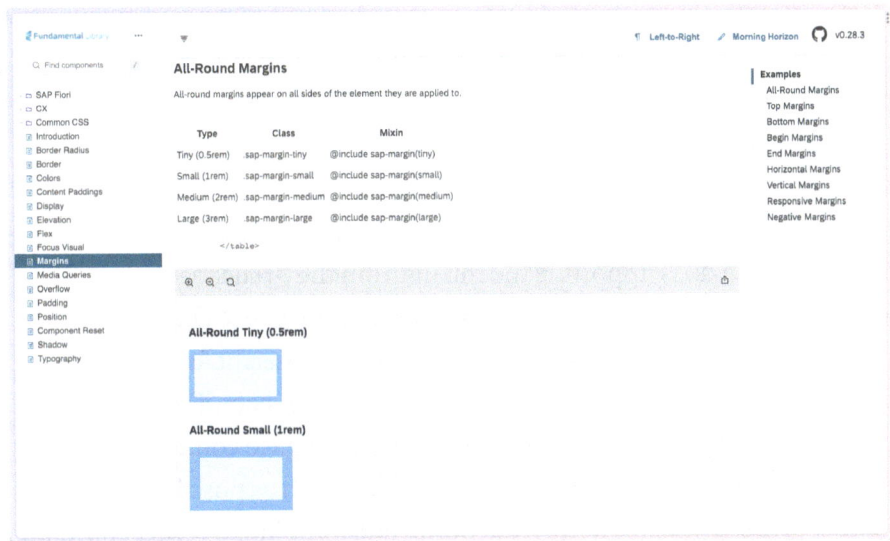

Figure 5-8. *The explanation of the* `@include sap-margin` *mixin*

Fundamental Library Styles provided the visual appearance of the new components we added to the application. However, the alignment components still might look a bit off. For example, the message strip is not hovering over the page, but it is actually displayed below it, and the form in the popover doesn't look very cleaned up either.

These problems can be addressed with the right style definitions, and, for this, we'll leverage Sass instead of CSS. As Sass is a superset of CSS, this can be done by **changing the filename** `style.css` to `style.scss`. At the same time, you need to **modify the import statement in main.js as well (Listing 5-17)**.

Listing 5-17. Import Sass to use with Vite

```
import "./style.scss"
```

By default, Vite doesn't handle Sass files. Luckily, these can be changed when you **add the development dependency in Listing 5-18**.

Listing 5-18. Add Sass as development dependency

```
npm add -D sass
```

A restart of the Vite development server will be necessary after this addition.

The rest of the Sass magic happens in this `style.scss` file. The first thing that we need to do is importing the mixins from the `@fundamental-styles/common-css` package. Then, we use the `sap-set-margin-all` and `sap-flex` mixins to turn the entire web application into a full-height flex box, with `flex-direction: column`, that has no margins. The second selector will make sure that both the page components and the page content use the `flex: auto` property to expand their height as much as possible. The third selector also uses mixins to turn the horizontal form into a vertical form that uses the same margin as in different places of the application while using a larger margin above the submit button to highlight it.

And lastly, the message strip selector defines rules for our two custom CSS classes to make sure the message strip can be hidden and floats above the page.

Replace everything, but the code defining the fonts, in `style.scss` with the code in Listing 5-19.

Listing 5-19. Utilize Sass to style the app

```
@import "@fundamental-styles/common-css/dist/sass/_common-
mixins.scss";

html,
body,
#root {
  height: 100%;
  @include sap-set-margin-all;
  @include sap-flex(column);
}
```

```scss
.fd-page {
  @include sap-flex-child(auto);

  & .fd-page__content {
    @include sap-flex-child(auto);
  }
}

.fd-popover__wrapper {
  @include sap-flex(column);
  @include sap-margin(tiny);

  & ui5-button {
    // @include sap-margin(small);
    @include sap-set-margin-top(1em);
  }
}

.fd-message-strip {
  &.exercise-float-bottom {
    @include sap-set-top-right-bottom-left(inherit, 1em,
    1em, 1em);
    @include sap-elevation(2, fixed);
  }

  &.exercise-hidden {
    opacity: 0;
    visibility: hidden;
    z-index: -1;
  }
}

@font-face {
  ...
```

The web application should look different once you saved this file. The page now covers the full height even after you cleared all list items, and the popover form shows the inputs above each other instead of next to each other. Compare your application to the one in Figure 5-9 and enjoy the improved style of your first application that uses Fundamental Library Styles.

Figure 5-9. *The final result of this exercise with a shortened list*

Build and Deployment

The build process doesn't need any further modification and can be triggered with the npm script in Listing 5-20.

Listing 5-20. Build the project with Vite

```
npm run build
```

Similar to the exercise of the previous chapter, it's up to the developer or architect to choose the appropriate tools and processes for debugging, testing, and deployment. All needed files will be generated in the output folder, dist/ by default.

The Road from Here

If you like this development approach, we recommend that you continue your learning path with a visit to the Storybook of the library.[12] This page contains some basic instructions on how you can get started with your next project and has a detailed list of all available components with usage instructions and code samples and their preview. Each component is introduced isolated from the others, so you can explore and compare them individually in different themes.

Most likely, you won't start a new project with vanilla JavaScript but instead already have made some decisions about frameworks you want to use, and maybe you already started the implementation with another component library. In this case, it makes sense to evaluate how Fundamental Library Styles can be integrated into your individual tech stack. This shouldn't be hard as we're talking about an npm package that includes style sheets, but the import might still work differently than the way we used earlier. If your stack uses Angular, we recommend having a look at the wrapper library[13] that comes with framework-native modules that also include the visual appearance as well as the behavior of the components. And if you're using other SPA frameworks, keep in mind that you can always combine Fundamental Library Styles (`@fundamental-styles/common-css`) with UI5 Web Components to leverage ready-to-use components in any client-side environment.

Similar to projects that include UI5 Web Components, this technology is just one cornerstone of your tech stack. It's up to you as a developer or architect to carefully curate which technologies, IDE extensions, and platforms you want to add. Further, it makes sense to dive deeper into the technologies and join the respective communities to make sure you follow the latest innovations and best practices there.

---

[12] https://sap.github.io/fundamental-styles/
[13] https://sap.github.io/fundamental-ngx/#/core/home

And remember, this is still a developing project that has a growing community. So connect with your peers, either in real life or online, and share your experiences and help shape the future of modern SAP web applications that use this technology.

Fundamental Library Styles for Enterprise Applications

This chapter showed you the advantages Fundamental Library Styles can offer when building enterprise software. With its base in CSS style sheets, this open source project brings the visual appearance of the SAP Fiori user experience to the most fundamental presentation layer of web technologies. And therefore, they can easily be added to existing projects without the burden of introducing new technical debt.

Similar to projects that use UI5 Web Components, this allows for the hiring of frontend engineers with different skill sets and also provides the freedom to hire UX engineers and designers who can use the CSS mixins to realize their styling visions. In fact, of all the technologies we discussed so far in this book, this one provides the most flexibility and freedom to developers. This is great if you, for example, find yourself in the tough spot where you need to add an SAP Fiori look and feel on top of an existing application. This can also be referred to as brownfield development.[14] But consequentially, you have to deal with all the requirements discussed in Chapter 1 (such as scalability, integration, i18n, performance, maintainability, and compatibility) and the resulting total cost of ownership. This means it might not be the best choice if you want to start a new traditional enterprise application project from scratch.

---

[14] https://en.wikipedia.org/wiki/Brownfield_(software_development)

Reimplementing "bread and butter" control with Fundamental Library Styles would be a lot of work, but this is also not what the technology has been built for. If you just want to use these controls in web projects with modern SPA frameworks, you can always fall back to UI5 Web Components that can work hand in hand with Fundamental Library Styles.

In the next chapter, we'll take a look at industry trends that influence SAP UI technologies.

CHAPTER 6

Industry Trends Impacting SAP UI Technologies

Chapter 1 already emphasized that consumer-grade technology has a big influence on enterprise software. In this chapter, we'll take a look at current trends coming from that side of the tech world. Each section of this chapter will introduce a common problem the industry faces today and briefly explain a trend that promises to address this problem, as well as show the challenges that come with this trend. The sections will also evaluate the relevance of that problem to enterprise software and particularly to the SAP UI universe and mention incubating projects from SAP to provide a standard solution. This chapter should be seen as a potential outlook in the future and inspiration for spare time projects, but not as a guaranteed forecast for upcoming technologies.

Visual UI Development

All businesses share the same problem: their resources are limited, and actions need to be prioritized. In enterprise IT, the impact of the mentioned situation is particularly pronounced, as highlighted in the introductory chapter. This is primarily due to the fact that the resources

© Marius Obert and Volker Buzek 2023
M. Obert and V. Buzek, *SAP UI Frameworks for Enterprise Developers*,
https://doi.org/10.1007/978-1-4842-9535-9_6

involved, specifically the people, tend to be more costly. The developers designing such systems need to be experts in software development as well as the business domain. While there are certainly many great interdisciplinary teams, they are not enough to satisfy the demand, which makes this resource even more scarce. The long process of gathering requirements, building a prototype, refining the solution, and deploying it at scale is just too expensive for many use cases.

Low code/no code (LCNC) describes approaches to build software applications that require minimal or no coding by the developer. Instead of writing code from scratch, developers can use prebuilt components, templates, and visual programming tools to create applications. Low-code development platforms provide a set of predefined building blocks, such as UI elements, business logic, and integration points, that developers can use to build applications quickly and easily. No-code development platforms take this concept even further by providing a fully visual interface for building applications. With no-code platforms, developers do not need to have any coding skills or knowledge of programming languages. Instead, they can use a series of visual tools and prebuilt templates to create and customize applications. LCNC tools typically provide a range of UI elements such as buttons, forms, lists, and navigation bars that developers can use to build the interface for their application. These elements can be customized using visual tools, such as drag-and-drop interfaces, property editors, and style sheets. The idea is that this simplification will not only allow faster development but also at a lower cost. These tools allow people without extensive software development knowledge, such as subject-matter experts, to create new solutions. This increases the number of employees who can implement these solutions, expanding the pool of potential contributors. With LCNC tools, these employees can focus on solving business problems rather than dealing with technical complexities, which helps to increase productivity and collaboration among team members. At the same time, these tools promise built-in security and compliance features. They say the applications created meet industry standards and regulations, making it easier for enterprises to comply with legal requirements. These platforms can be particularly useful in situations where time is of the essence or when the development team is small and needs to move quickly. However, they may not be suitable for all types of projects, and it is important for developers to carefully evaluate whether these approaches are appropriate for their specific needs.

LCNC platforms are often used to create prototypes, MVPs (minimum viable products), and small-scale applications quickly and with minimal up-front investment.

LCNC platforms promise numerous benefits. However, they also come with their own set of challenges. One challenge is the limited customization options offered by LCNC platforms. While these platforms provide prebuilt components and templates for rapid development, they may restrict the ability to address complex or unique business requirements and integrate with external systems. Another challenge is scalability and performance. As applications built with LCNC grow in complexity and user load, they may encounter limitations in scaling and optimizing performance. Abstracting away technical complexities can hinder the ability to fine-tune applications for higher demands.

Vendor lock-in is also a concern. Adopting a specific LCNC platform can create dependence on proprietary tools and infrastructure, making it challenging to switch platforms or migrate custom code. This lack of flexibility can have long-term implications. Security and compliance are important considerations. While LCNC platforms provide security features, customization and control over security measures may be limited. Organizations must ensure that these tools meet their specific security and compliance requirements, especially for sensitive applications or industries with strict regulations. Technical limitations are another challenge. Complex integrations, advanced data processing, and custom business logic may exceed the capabilities of LCNC platforms. Projects requiring intricate algorithms or specialized workflows may still require skilled developers with advanced coding expertise. Lastly, there is a learning curve and skill set consideration. While LCNC platforms aim to empower employees, there is still a learning curve in understanding the platform's capabilities and configuring applications effectively. Skilled developers may still be needed for more complex projects or advanced customization. Considering these challenges is crucial when evaluating the adoption of low-code or no-code tools, ensuring they align with specific needs, scalability requirements, and long-term goals of the organization or project.

SAP identified this trend already a few years ago and acquired AppGyver in 2021. The company built a no-code development environment to empower business users and developers to visually create user interfaces without writing code. By now, SAP integrated this solution into its Business Technology Platform and rebranded it to SAP Build Apps.

The online tool offers multiple views to define the view, data model, theme, and other aspects of web applications. Figure 6-1 shows the UI canvas of the SAP Build Apps platform during the design phase of a user interface.

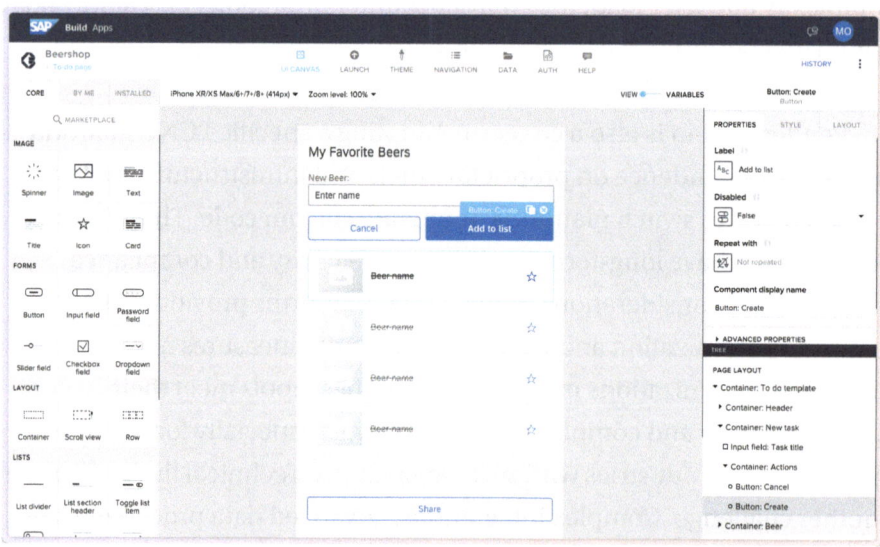

Figure 6-1. *The UI canvas of SAP Build Apps while building a list view*

While SAP Build Apps can be used to design new interfaces from scratch, SAP has another LC feature that is already baked-in in most SAPUI5 applications: SAPUI5 flexibility. SAPUI5 flexibility allows users and developers to customize the user interface of SAPUI5-based apps without modifying the source code. This feature addresses the need for tailored user experiences and enables different user groups to shape the UI according to their specific requirements. Changes made to the UI are separate from the source code, ensuring compatibility with future app upgrades. Apps developed with SAPUI5 typically support this feature as long as the respective backend technology supports storage of these customizations. The availability of SAPUI5 flexibility varies across platforms and SAP product suites, such as SAP NetWeaver; SAP S/4HANA Cloud; SAP BTP, Neo environment; SAP BTP, ABAP environment; and SAP Build Work Zone, standard edition.

SAPUI5 flexibility describes three different concepts: end-user personalization, key user adaptation, and developer adaptation. All of which are related to customizing the user interface of SAPUI5-based apps. However, they differ in terms of the roles involved and the extent of customization allowed. Table 6-1 shows a comparison of the three concepts.

Table 6-1. *Comparing end-user personalization, key user adaptation, and developer adaptation*

	End-User Personalization	Key User Adaptation	Developer Adaptation
Role	End users of the app	Business experts or key users responsible for coordinating teams	App developers or developers with technical expertise
Purpose	Allows individual users to personalize the UI according to their preferences and work needs	Enables key users to adapt the UI of an app to meet the specific needs of their teams	Involves preparing an app for UI changes or adapting the UI of an existing app for broader user groups
Customization options	Users can save filter settings, add frequently used links, modify sections, and arrange cards on overview pages	Key users can make changes such as moving UI elements, adding/removing elements, renaming elements, combining/splitting fields, changing settings, and embedding external content	Developers can use SAPUI5 flexibility features to extend SAPUI5 freestyle or SAP Fiori elements apps. Changes may include moving UI elements, adding/removing elements, changing properties, embedding external content, creating control variants, and more
Scope	Personalization is limited to the preferences of individual users and doesn't affect the app's core functionality or structure	Adaptation is typically done using SAP's development environment, and changes are made separate from the app's source code. App variants can be created for different user groups, but the original app remains unchanged	Developers work within SAP's development environment to create adaptation projects, make UI changes using tools like the SAPUI5 Visual Editor, and deploy app variants while keeping the original app intact

In summary, end-user personalization focuses on individual user preferences, key user adaptation (Figure 6-2) allows business experts to customize the UI for their teams, and developer adaptation involves preparing or modifying apps for UI changes on a broader scale. Each approach caters to different levels of customization and involves different roles in the process.

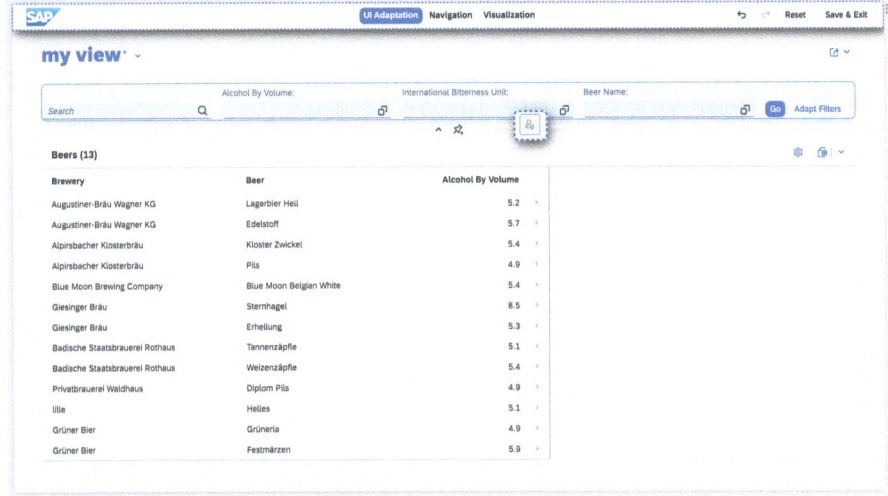

Figure 6-2. *UI offering key user adaptation for the SAP Fiori elements app from Chapter 3*

Type Safety in the Frontend

As you have certainly noticed, we've been using vanilla JavaScript for the examples throughout this book. This was intentional: we tried to keep the code as brief and simple as possible, to be run without any abstraction layers in between.

This is what TypeScript would have induced: an additional layer of abstraction. Loosely speaking, TypeScript is "JavaScript with types" plus syntactic sugar. It introduces static types that are denoted in the source code (Listing 6-1). This typing already helps to prevent accidental retypings of variables – in Listing 6-1, the TypeScript compiler denotes that message is supposed to be a String and can't be assigned a boolean. This would absolutely be possible and even intended in JavaScript: dynamic types and value assignments are an explicit feature of the language.

Listing 6-1. Static type example in TypeScript

```
let message: String = "hi!"

// Type 'boolean' is not assignable to type 'String'.
message = false //> possible in JavaScript!

// This expression is not callable.
// Type 'String' has no call signatures.
message()
```

While for simple variables this might look like unnecessary overhead, complex variables that are typed properly enable early error detection. TypeScript will emit an error pointing to the root cause if typed complex variables are not used properly anywhere in the codebase (Listing 6-2).

Listing 6-2. Complex type with helpful error messages upon wrong usage

```
const drink: MintJulep = {
    mint: {
        leaves: {
            min: 5,
            max: 8
        }
    },
    syrup: 0.5,
    bourbon: 2,
    ice: "cubes"
```

```
// error: Type '"cubes"' is not assignable to type '"crushed" | "fine" | "coarse"'.
}

export interface MintJulep {
    mint:    Mint;
    syrup:   number; // ounces
    bourbon: number; // ounces
    ice:     "crushed" | "fine" | "coarse";
}

interface Mint {
    leaves: Leaves;
    spring?: number; // for garnish
}

interface Leaves {
    min: number;
    max: number;
}
```

This helps to maintain large codebases in terms of making development less error and aids teams in using each other's code modules properly.

As already mentioned, TypeScript needs a build step to be transpiled to plain JavaScript that can in turn be executed by either the browser or Node.js. While this creates more effort for the developer, it can at the same time be put to good use. The transpilation result can be adjusted at will, also in parallel streams. For example, a deployment target might require JavaScript in a certain maximum version, like ECMAScript[1] 2022, while

---

[1] https://en.wikipedia.org/wiki/ECMAScript

another deployment target requires JavaScript as ECMAScript Modules
only, not CommonJS.[2] Both code requirements can be fulfilled in parallel
transpilation runs from TypeScript, albeit with different configurations
(Listing 6-3). This accelerates Continuous Deployment scenarios
tremendously and retains the TypeScript codebase as the single source
of truth.

Listing 6-3. Excerpt from TypeScript configurations for
transpilation

```
// tsconfig.json excerpt for ES Modules
{
    "compilerOptions": {
        "module": "ES2022",
        // ...
    }
    // ...
}

// tsconfig.json excerpt for way backwards-compatible JavaScript code
{
    "compilerOptions": {
        "module": "commonjs",
        "target": "ES5",
        // ...
    }
    // ...
}
```

---

[2] https://en.wikipedia.org/wiki/CommonJS

UI5 TypeScript is using Babel for the transpilation; the information from the `tsconfig.json` is relevant for the dev time only (VS Code). Everything around the actual transpile step has to go into a `.babelrc`, and you can use the `.browserslist` configuration to define the target browser environment.

Additionally, TypeScript unlocks certain programming features dearly missed in vanilla JavaScript. One example is overloading parameters of a function. In plain JavaScript, defining the same method with different parameters is not possible (Listing 6-4).

Listing 6-4. Method overloading is not possible in JavaScript

```
// Radler
function tap(beer, lemonade = "citrus") {
  return {
    beer: 0.5,
    [lemonade]: 0.5
  }
}

// this "overwrites" aka redefines tap()
function tap(beer) {
  return { [beer]: 1 }
}

// both return
// { Helles: 1 }
tap("Helles") // intended: { Helles: 1 }
tap("Helles", "citrus") // intended: { beer: 0.5, citrus: 0.5 }
```

With TypeScript, method overloading is achievable, as it is a core feature of the language abstraction (Listing 6-5). Providing different signatures for the same implementation is a matter of listing the various call possibilities of a method.

Listing 6-5. Method overloading in TypeScript

```
function tap(beer: string): number
function tap(beer: string, lemonade: "citrus"): number
// Radler
function tap(beer: string, lemonade: "citrus", flavour:
string): number

function tap(beer: string, lemonade?: string, flavour?:
string): number {
  if (lemonade && lemonade !== "citrus") {
    throw new Error("never mix a Radler with anything but
    citrus")
  }
  if (lemonade && beer !== "Helles") {
    throw new Error("never mix a Radler with anything but a
    Helles")
  }
  if( lemonade && flavour ) {
    throw new Error("don't do this. just don't.")
  }
  // ...
  return 1
}
```

```
tap("Pilsner") // 1
tap("Pilsner", "citrus") // Error: never mix a Radler with
anything but a Helles
tap("Helles", "citrus", "grenadine") // Error: don't do this.
just don't.
```

When applied to UI5, TypeScript allows for a much cleaner syntax of controllers and other business logic coding.

In Listing 6-6, you can see the same UI5 controller side by side, once in JavaScript and once in TypeScript. The TypeScript version looks not only cleaner but is also semantically more on point. It extends the Controller module into a dedicated class that is subsequently exported. Also, the proprietary sap.ui.define is omitted, reducing notation complexity.

Listing 6-6. A JavaScript and a TypeScript UI5 controller side by side

```
// JavaScript
sap.ui.define(
  ["sap/ui/core/mvc/Controller"],
  /**
   * @param {typeof sap.ui.core.mvc.Controller} Controller
   */
  function (Controller) {
    "use strict"

    return Controller.extend("com.apress.openui5.
    controller.App", {
      onInit() {}
    })
  }
)
```

```
// TypeScript
import Controller from "sap/ui/core/mvc/Controller"

/**
 * @namespace com.apress.openui5.controller
 */
export default class App extends Controller {
  public onInit(): void {}
}
```

TypeScript, of course, also fulfills its initial purpose to introduce type checking at development time. Additionally, navigating a codebase by clicking through variables and module names becomes easier thanks to the provided UI5 type[3] definitions. Add the cleaner syntax, and the overall developer experience is tremendously increased – that is why many people, including the authors of this book, regard writing UI5 applications in TypeScript the next iteration of the framework.

As with all things, the upsides come with challenges or even downsides as well.

For one, the UI5 TypeScript applications need an additional step before they can be run, tested, or deployed. The open source babel module transform-ui5[4] is designed for that transpilation of UI5 apps – during development time as well as to build the project for deployment. It is used in the ui5-tooling-transpile npm[5] module, which in turn plugs into the UI5 Tooling server. The server then automagically transpiles TypeScript files in the webapp folder. The same ui5-tooling-transpile

---

[3] www.npmjs.com/package/@openui5/types

[4] https://github.com/ui5-community/babel-plugin-transform-modules-ui5

[5] https://github.com/ui5-community/ui5-ecosystem-showcase/tree/main/packages/ui5-tooling-transpile

npm module[6] can be used with the UI5 Tooling builder as a preparation step for deployment and even sharing the same babel configuration. A very detailed example with thorough explanations on how to develop a UI5 application with TypeScript lives in the SAP Samples GitHub repository.[7]

The transpiled JavaScript files are intended for machine execution, either by a browser or by Node.js. They require quite an effort to be read by humans. Listing 6-7 shows above minimal UI5 controller written in TypeScript when transpiled to JavaScript with default TypeScript settings.

Listing 6-7. Transpilation of the minimal UI5 TypeScript controller into JavaScript

```
"use strict";
var __extends = (this && this.__extends) || (function () {
    var extendStatics = function (d, b) {
        extendStatics = Object.setPrototypeOf ||
            ({ __proto__: [] } instanceof Array && function
            (d, b) { d.__proto__ = b; }) ||
            function (d, b) { for (var p in b) if (Object.
            prototype.hasOwnProperty.call(b, p))
            d[p] = b[p]; };
        return extendStatics(d, b);
    };
```

---

[6] https://github.com/ui5-community/ui5-ecosystem-showcase/tree/main/packages/ui5-tooling-transpile

[7] https://github.com/SAP-samples/ui5-typescript-tutorial

```
        return function (d, b) {
            if (typeof b !== "function" && b !== null)
                throw new TypeError("Class extends value " +
                String(b) + " is not a constructor or null");
            extendStatics(d, b);
            function __() { this.constructor = d; }
            d.prototype = b === null ? Object.create(b) :
            (__.prototype = b.prototype, new __());
        };
    })();
Object.defineProperty(exports, "__esModule", { value: true });
var Controller_1 = require("sap/ui/core/mvc/Controller");
/**
 * @namespace com.apress.openui5.controller
 */
var App = /** @class */ (function (_super) {
    __extends(App, _super);
    function App() {
        return _super !== null && _super.apply(this, arguments)
        || this;
    }
    App.prototype.onInit = function () { };
    return App;
}(Controller_1.default));
exports.default = App;
```

This makes debugging the UI5 application more difficult, as the JavaScript sources look convoluted. Fortunately, the transpilation step can also omit source maps. Those can be considered a glue between JavaScript and TypeScript parts. Mozilla[8] says: "A source map is a file that maps from the transformed source to the original source, enabling the browser to reconstruct the original source and present the reconstructed original in the debugger." By using source maps, UI5 TypeScript applications can be comfortably debugged again – ui5-tooling-transpile will emit source maps in line with the generated JavaScript sources.

When using UI5 TypeScript, the code in Listing 6-7 is just the intermediate build result when transpiling the TypeScript code to JavaScript code. As explained earlier, UI5 relies on Babel for the entire transpilation process, not on the TypeScript compiler tsc. If the UI5 Babel transpilation steps are applied to the TypeScript controller from Listing 6-6, it omits a very readable JavaScript version (Listing 6-8).

Listing 6-8. Babel-based transpilation of UI5 TypeScript controller from Listing 6-6

```
"use strict";
sap.ui.define(["sap/ui/core/mvc/Controller"], function
(Controller) {
  /**
   * @namespace com.apress.openui5.controller
   */
  var App = Controller.extend("com.apress.openui5.
  controller.App", {
    onInit: function _onInit() {}
  });
  return App;
});
```

---

[8] https://firefox-source-docs.mozilla.org/devtools-user/debugger/
how_to/use_a_source_map/index.html

Looking at the test stack of Chapter 2 reveals a gap for UI5 TypeScript. While using QUnit with TypeScript[9] for unit testing is straightforward, and wdi5 supports end-to-end tests written in TypeScript[10] out of the box, using OPA5 with TypeScript for integration tests still has its challenges. For the most part, this is due to the yet missing type definitions for the OPA5 library.

But nevertheless, SAP's take on this is clear: even though officially still flagged as beta at the time of this writing, TypeScript is the future for UI5 application development. There's a dedicated repository with tips and tricks[11] for UI5 apps with TypeScript. And the easy-ui5 application generator we used in Chapter 2 is already equipped and maintained with TypeScript skeletons[12] as well.

Micro Frontends

Traditional frontend architectures, from now on referred to as monolithic frontend architectures, face several challenges. They become complex and difficult to maintain as applications grow in complexity. This could lead to interteam dependencies hindering team autonomy and decision-making and eventually slowing down development. Another potential issue is that any change or update requires the entire application to be rebuilt and redeployed, causing bottlenecks and slower release cycles. Monolithic architectures also lack the flexibility to accommodate diverse frontend

---

[9] https://github.com/SAP-samples/ui5-typescript-helloworld/tree/testing

[10] https://github.com/ui5-community/wdi5/tree/main/examples/ui5-ts-app/test/e2e

[11] https://sap.github.io/ui5-typescript/

[12] https://github.com/ui5-community?q=generator+ts&type=all&language=&sort=

technologies and frameworks used by different teams or legacy systems – whether this might or might not be beneficial. And finally, failures in one part of the monolith can impact the entire application, making it less resilient and stable. And the rising architecture pattern "Micro Frontends" aims to solve these problems by offering modularity, independent development and deployment, technology diversity, team autonomy, and improved resilience and speed. Micro Frontends describe a pattern in which a web application is built as a collection of small, independent, and self-contained components, each of which represents a Micro Frontend – analogous to the microservices pattern in the backend. These components are typically developed and maintained by separate teams, and they are deployed and served independently of one another. This can help to improve the overall development process by allowing teams to work on their own components without having to coordinate with other teams or deal with conflicts in a shared codebase. It can also make it easier to scale the frontend of a web application, as new Micro Frontends can be added or removed without having to adjust the entire application.

Micro Frontend architecture, despite its benefits, also presents certain challenges. One challenge is the increased complexity compared to traditional monolithic architectures. Managing multiple independent frontend applications and their interactions requires careful coordination and communication between teams. It can be challenging to ensure consistency in design, user experience, and behavior across different Micro Frontends. Another challenge is the deployment and infrastructure overhead. With multiple frontend applications to deploy and manage, there is an increased deployment and infrastructure workload. Each Micro Frontend needs to be deployed separately, potentially leading to more complex deployment pipelines and increased infrastructure requirements. Coordinating deployments and handling cross-cutting concerns like authentication and routing can be demanding. Inter-Micro Frontend communication is another challenge, as they often need to communicate with each other to provide a seamless user experience. These functionalities need to be handled consistently across different Micro Frontends. A Micro Frontend architecture can also introduce additional overhead in terms of network requests and loading times. Coordinating and optimizing the loading of different Micro Frontends, especially in scenarios where multiple Micro Frontends are required for a single page, is essential to ensure good performance and minimize latency.

Testing and quality assurance also become more challenging in a Micro Frontend architecture. Overall, addressing these challenges effectively requires establishing best practices, adopting suitable tools and frameworks.

While many experts agree on the general properties of a Micro Frontend, there is (so far) no consent on the exact definition. So, there might be voices who say the SAP Fiori launchpad, as we got to know it in this book, is a Micro Frontend. The SAP Fiori launchpad follows a modular structure, where different tiles and apps can be developed independently by separate teams. Each tile or app can be considered a Micro Frontend, as it operates as a self-contained unit within the launchpad, and each app can be based on a different version of SAPUI5. Opponents of this classification could argue the SAP Fiori launchpad serves as a single entry point for accessing various applications and tiles, and the launchpad itself acts as a monolithic container for these Micro Frontends, centralizing the user interface and navigation while restricting other technologies than SAPUI5.

Project Luigi[13] is an open source JavaScript framework from SAP for implementing Micro Frontends. It enables the breakdown of large frontend monoliths into smaller, more manageable chunks that can be developed by independent teams. The framework is already being used in production and close-to-production within SAP and outside of

---

[13] https://docs.luigi-project.io/

SAP, for example, in Kyma, SAP C/4HANA Cockpit and Varkes. It offers configuration options, API functions, and built-in features that facilitate the migration to a Micro Frontend architecture. Luigi enables a consistent user navigation experience across Micro Frontends, regardless of the underlying technology stack – such as Angular, React, Vue, or UI5. Luigi consists of two main components: Luigi Core and Luigi Client. Luigi Core serves as the main app where Micro Frontends are embedded, offering features like navigation, authorization, localization, and general settings configuration. Luigi Client, on the other hand, provides API functions and communication capabilities for each of the Micro Frontends.

To get started with Luigi, you can set up a Luigi Core application on your preferred framework or use the provided options like Angular, UI5, Vue, or React. Additionally, you can add Luigi Client features to your existing Micro Frontends by installing Luigi Client and leveraging its API functions. The Luigi Fiddle website[14] shows an interactive live demo of a Luigi Micro Frontend that combines the technologies this book introduced: UI5, UI5 Web Components, and Fundamental Library Styles. Figure 6-3 also shows the general structure of a Luigi app that follows the SAP Fiori guideline of having a shell bar on top. It also uses a sidebar on the left-hand side to switch between the Micro Frontends and the main view, which lists the available Micro Frontends.

---

[14] https://fiddle.luigi-project.io/#/home/overview

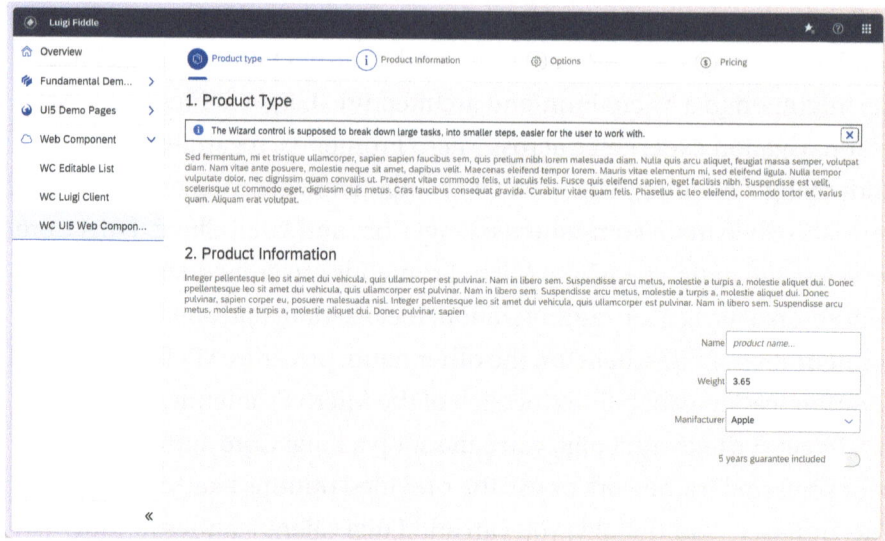

Figure 6-3. *Interactive Luigi Fiddle*

GraphQL

When building web applications with plain HTTP request, developers often face challenges in efficiently fetching and managing data, which can lead to performance bottlenecks, increased network overhead, and unnecessary processing. Without a solution, developers might retrieve excessive or insufficient data, a.k.a. overfetching or underfetching data, impacting application efficiency. Fetching related data requires multiple round trips to the server, resulting in slower response times and increased latency. Frontend and backend collaboration can be hindered as frontend developers rely on backend changes for their data needs, leading to delays in development. These issues hinder application scalability and flexibility, limiting the ability to accommodate evolving client demands. By addressing these problems, a solution empowers developers to request precisely the required data, reduce round trips, optimize performance, and streamline development efforts.

With OData (see Chapter 3), the enterprise software world already addressed this problem. However, GraphQL[15] is an alternative solution that is currently very popular among developers of consumer-grade apps. GraphQL is a query language that was developed by Facebook in 2012 and released as an open source project in 2015. It is widely used in modern web and mobile development and has a large and active community of developers and users. To use GraphQL, developers define a schema for their API that specifies the types of data that are available, as well as the relationships between those types. This schema is written in the GraphQL schema definition language. Clients can then send queries to the API using this schema as a guide, requesting specific data and relationships. The server will execute the query and return the requested data in a JSON format.

Listing 6-9 shows the payload of a request querying for a list of beers. For each beer, we want to retrieve its name, alcohol by volume, international bitterness units, and information about the brewery including its name.

Listing 6-9. GraphQL sample request

```
query {
  beers {
    name
    abv
    ibu
    brewery {
      name
    }
  }
}
```

---

[15] https://graphql.org/

Listing 6-10 shows the matching response. Each beer object includes its name, abv, ibu, and the associated brewery object with its name. The response provides a structured and organized representation of the requested data, making it easy to work with and display the information in the client application without over- or underfetching.

Listing 6-10. GraphQL sample response

```
{
  "data": {
    "beers": [
      {
        "name": "Erhellung",
        "abv": "5.3",
        "ibu": "18",
        "brewery": {
          "name": "Giesinger Bräu"
        }
      },
      {
        "name": "Grünerla",
        "abv": "4.9",
        "ibu": "11",
        "brewery": {
          "name": "Grüner Bier"
        }
      }
    ]
  }
}
```

GraphQL and OData are both protocols that are used to expose and query data over HTTP. OData is often used in enterprise scenarios where there is a need to expose a well-defined data model to clients, while GraphQL is more commonly used in situations where the client needs more control over the data it retrieves and the ability to ask for specific subsets of data. But as we learned in Chapter 1, these lines start to get blurry. So let's focus on some other key differences between these protocols:

1. *Query language*: GraphQL uses a custom query language that is based on the structure of the data, while OData uses a standard query language based on the OData specification.

2. *Data model*: OData's fixed data model defines the structure of the data and the relationships between different entities. GraphQL, on the other hand, is more flexible and does not have a fixed data model. Instead, it allows the client to specify the specific data that it needs, and the server responds with the requested data.

3. *Versioning*: OData APIs are versioned using a fixed schema that defines the structure of the data. This means that any changes to the API must be made in a new version, which can be difficult to manage in a large enterprise. In contrast, GraphQL allows the server to evolve the API over time without breaking existing clients, as long as it maintains backward compatibility.

4. *Performance*: OData is generally more efficient when it comes to retrieving large amounts of data, as it allows the client to specify which fields it intends to retrieve and supports server-side filtering and paging. GraphQL requires the server to execute a separate resolver function for each field, which can be slower for complex queries.

In summary, GraphQL offers more flexibility and control over the data that is returned, and it is better suited for environments where the data model is constantly evolving. OData, on the other hand, is more suited for environments where the data model is more stable and where it is important to have a fixed schema for the data. This blog post[16] by DJ Adams provides a more detailed comparison of both protocols.

SAP is, rightfully, still leveraging OData for most of their standard applications and will do so for the foreseeable future. At the same time, they recognized that many community members want to embrace GraphQL as well. This holds especially for newer web frameworks that could be *fiorified with* UI5 Web Components or Fundamental Library Styles. There has also been an open source[17] proof of concept showing how GraphQL could be integrated with OpenUI5 at UI5Con on Air 2020.[18] UI5 comes with a library called UI Integration Cards.[19] These integration cards provide a way to embed external content, such as data and services, into

---

[16] https://blogs.sap.com/2018/09/03/monday-morning-thoughts-considering-graphql/

[17] https://github.com/petermuessig/ui5-sample-apollo

[18] www.youtube.com/watch?v=r1XChmnI5gw

[19] https://sapui5.hana.ondemand.com/test-resources/sap/ui/integration/demokit/cardExplorer/webapp/index.html#/overview/introduction

any kind of web application. They allow developers to create reusable and customizable card components that can be easily integrated into any web application. These Integration Cards provide a consistent and flexible way to display data from various sources, one of which are GraphQL sources as shown in Figure 6-4.

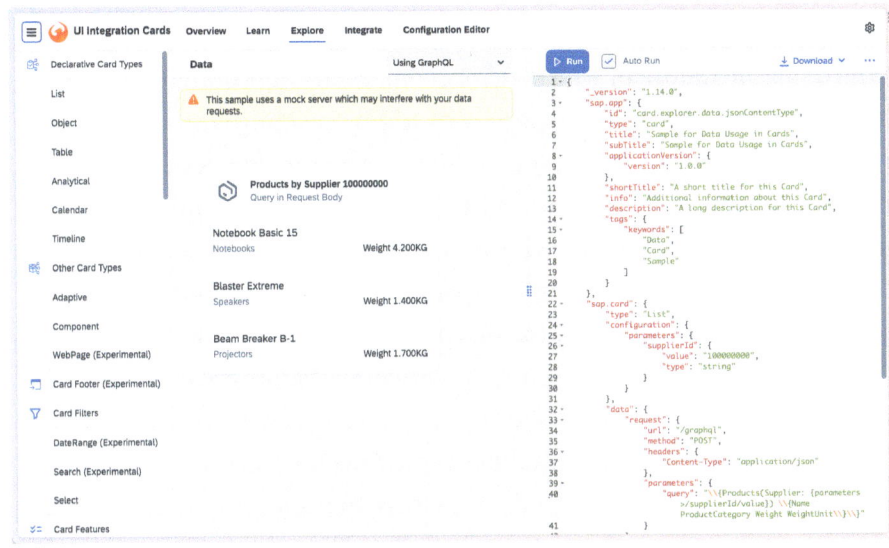

Figure 6-4. *Demo and exploration page for UI Integration Cards[20]*

At the same time, the SAP Cloud Application Programming Model (CAP), SAP's modern framework for backend developers, is also open to other protocols. The framework offers the option to leverage adapters[21] to include community-driven projects,[22] one of which can generate a GraphQL schema for the data model and serve an endpoint using GraphQL.

---

[20] https://sapui5.hana.ondemand.com/test-resources/sap/ui/integration/demokit/cardExplorer/webapp/index.html#/explore/data/graphql

[21] https://cap.cloud.sap/docs/node.js/protocols?q=graphql#graphql-adapter

[22] https://github.com/cap-js/graphql

Next-Gen Full-Stack Web Frameworks

Continuous innovation and the constant introduction of new technologies drive the ever-changing landscape of web development. Developers are presented with frequent releases of programming languages, frameworks, libraries, and tools that offer enhanced approaches to building web applications. Additionally, the evolving expectations of users contribute to the dynamic nature of web development. Users now demand fast, responsive, and immersive experiences when engaging with web applications. To fulfill these expectations, developers must embrace novel techniques and frameworks that enable them to create modern and captivating user experiences. This demand for improved user experiences fuels the development of (new and existing) frameworks and tools that cater to evolving needs, resulting in a dynamic landscape as developers strive to meet these requirements. Framework creators are consistently enhancing and refining their offerings to gain a competitive advantage. The resulting competition stimulates rapid development and release of new frameworks, each equipped with distinctive features, performance optimizations, and developer-friendly tools.

The next generation of full-stack web frameworks is a product of this continuous innovation and evolution within the web development landscape. These frameworks, such as Next.js,[23] Nuxt.js,[24] Nest.js,[25] and SvelteKit,[26] represent the latest generation of tools that aim to address the evolving needs of developers and users.

---

[23] https://nextjs.org/

[24] https://nuxtjs.org/

[25] https://nestjs.com/

[26] https://kit.svelte.dev/

These web frameworks incorporate the latest advancements in technology and development practices. They often provide features like server-side rendering (SSR), static site generation (SSG), code splitting, and efficient data fetching. These frameworks prioritize performance, scalability, and developer productivity while delivering fast and immersive user experiences.

Server-side rendering is a technique used in web development to render a web page on the server instead of the client. This can be useful in a number of situations, such as when the client's device lacks the processing power or network bandwidth to render the page efficiently or when the page contains sensitive information that should not be transmitted to the client.

There are several frameworks available that typically work by providing a set of tools and libraries that allow developers to write code that runs on the server, generates the HTML for a web page, and then sends that HTML back to the client for rendering in the user's web browser.

These web frameworks promise improved developer productivity by providing streamlined workflows, reducing boilerplate code, and offering intuitive APIs. They prioritize performance optimization by rendering pages on the server and optimizing asset loading; they deliver faster initial load times, improved user experience, and better search engine visibility. These frameworks aim for scalability and robustness by providing architectural patterns, built-in routing, state management, and data handling capabilities that support building complex and scalable applications.

At the same time, the learning curve of these next-gen frameworks can be a challenge. Developers need to invest time in understanding the concepts, APIs, and configuration options specific to each framework. Secondly, integration challenges may arise when incorporating them with existing systems, legacy codebases, or third-party libraries. Compatibility issues, configuration conflicts, or unfamiliarity with integration patterns can hinder the development process and require additional effort. And with their feature-rich nature, they may introduce unnecessary complexity for simpler projects. Improper configuration or misuse of certain features can even have a negative performance impact. Developers need to understand the best practices and performance optimizations to ensure efficient usage of these frameworks and maintain high-performance standards.

SAP also recognized the need for a modern full-stack technology stack that integrates frontend and backend seamlessly. And while SAP doesn't provide a single framework for this job, they offer a palette of tools that harmonizes well. For the frontend part with UI5 and SAP Fiori elements, we want to refer to Chapters 2 and 3 of this book. And the framework of choice to implement the backend of modern enterprise applications is the SAP Cloud Application Programming Model and the corresponding development tools.[27] While there isn't a one-click installation for these tools, they are still tightly coupled. As Figure 6-5 shows, the wizard to create an SAP Fiori elements app can be based on the data model of a local CAP backend application.

---

[27] https://cap.cloud.sap/docs/tools/

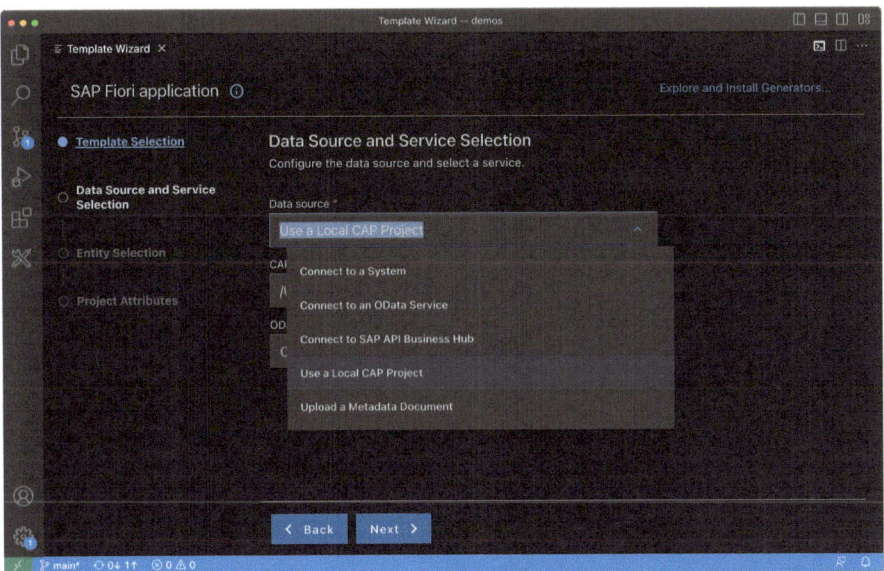

Figure 6-5. *The SAP Fiori tools wizard offering to connect to a local CAP project*

It is also possible to make use of the before mentioned next-gen full-stack web framework and leverage new rendering strategies such as SSG and SSR for enterprise applications. Developers can follow the best practices defined in the developer resources section of the respective framework of choice and build the UI components with Fundamental Library Styles (see Chapter 5). If you would rather like to use UI5 Web Components, you might need to be a bit patient. At the time of writing, the UI5 Web Components for React still had some dependencies on browser APIs and, therefore, couldn't be rendered solely on the server. However, SAP is aware of this shortcoming[28] and working on this, which means we can expect that this feature will be available eventually.

---

[28] https://github.com/SAP/ui5-webcomponents-react/issues/4091

It is important that developers understand these concepts and carefully evaluate whether their enterprise application benefits from the used rendering strategies. After all, there has been a deliberate decision by SAP to move away from server-rendered user interfaces (like Web Dynpro and SAP GUI) when introducing the first version of SAPUI5. Enterprise applications often involve complex business logic, dynamic content, and personalized experiences. Implementing SSR or SSG in such cases can introduce additional complexity and challenges. Enterprise applications often rely on real-time data updates, such as product information or purchase orders. SSR and SSG are better suited for static or semistatic content where frequent updates are not required. Implementing real-time updates with SSR or SSG can be more complex and less efficient compared to using client-side rendering or WebSocket-based solutions.

Enterprise applications frequently deal with sensitive data and have stringent security requirements. SSR and SSG can introduce additional security considerations, such as handling server-side data access, securing data transfers, and managing user authentication and authorization. These security complexities need to be carefully addressed, which can add overhead to the development and maintenance process. And finally, many next-gen frameworks require server infrastructure or specialized hosting setups that support SSG and SSR. In some enterprise contexts, there may be limitations or restrictions on the choice of hosting platforms or infrastructure. This can add complexity to the deployment and maintenance process, requiring coordination with IT teams or adherence to specific hosting policies.

## CHAPTER 7

Staying on Top of the Ball

The previous chapters introduced all the current web technologies in the SAP universe and beyond. Now it's time for you to continue the learning journey and deepen your knowledge. There are multiple ways you can do this. Same as in other areas of software development, self-studying with resources on the Web and learning on the job is a big aspect. But it's understandably hard to do this on your own, especially if you are not yet working with these technologies in your day job. So, this chapter will show you communities of like-minded learners who are interested in the same technologies and willing to share their knowledge.

Meeting the Community in Real Life

There are multiple reasons that speak for in-person events to learn about new technologies and develop your skills. The most obvious is that in-person events allow developers to learn about new technologies, best practices, and industry trends. This can be especially helpful if you "accidentally" learn about a new topic that wasn't on your radar. Another benefit is that these events provide a chance to hear from recognized experts and thought leaders in the enterprise technology field, which can be inspiring and help young and old developers think about their work in

© Marius Obert and Volker Buzek 2023
M. Obert and V. Buzek, *SAP UI Frameworks for Enterprise Developers*,
https://doi.org/10.1007/978-1-4842-9535-9_7

new ways. And finally, social events are a fun and exciting way to spend a few days and learn casually – and allow attendees to relax and have a good time. While this section focuses on in-person events, it also includes a few virtual events that had to shift to a virtual setting during the COVID-19 pandemic and now continue as a hybrid offering consisting of in-person and virtual aspects.

Depending on your region, there might be more or less in-person events. But that doesn't mean this section is not for you. Maybe you can start by connecting virtually to a local community in your area or start one with friends. It's probably easier to build virtual asynchronous events where everyone can join at their favorite time. And then, over time, you can grow into synchronous events and eventually plan your own in-person community event or meetup.

SAP TechEd

The largest SAP-related technology conference is SAP TechEd.[1] This multiday event is organized by SAP and attracts developers, IT professionals, and business users eager to learn about SAP's products and technologies. The event covers various topics beyond pure user interface technologies, including data management, analytics, cloud computing, and artificial intelligence.

The conference typically consists of keynotes, breakout sessions, hands-on workshops, demonstrations, and an expo area – all focused on helping attendees learn how to use SAP technology effectively to drive business value. In addition to the technical content, SAP TechEd also provides attendees with networking opportunities, allowing them to connect with other SAP customers, partners, and experts worldwide.

---

[1] www.sap.com/germany/about/events/teched.html

Before the COVID-19 pandemic, the event charged a fee for the passes that took place in three cities across the globe: in Las Vegas, for the North and South American market; in Barcelona, for Europe, the Middle East, and Africa; and in Bangalore, for the Asian market. During the pandemic, the event transitioned temporarily to a fully virtual event that live-streamed all sessions for free.

SAP User Group Events

SAP User Groups are independent, nonprofit organizations representing the interests of SAP customers, partners, and users in their region. These organizations offer a range of resources and services to their members.

They are committed to helping their members get the most out of their SAP investments by providing opportunities to learn, connect, and collaborate with other SAP users. This includes hosting regular meetings, educational events, peer-to-peer networking, and other events and providing online resources and support through their websites and member portals. The focus area of these events can be very broad or focused on a specific technology area, and the speakers are, in most cases, either SAP employees or user group members.

In addition to providing support and resources to their members, the user groups also work closely with SAP to advocate for the needs and interests of their members. This includes representing the voice of their members in SAP product development and influencing SAP's strategic direction.

Each organization has a website on which they promote their upcoming events. While some events are only available to members, others are open to everyone.

The Deutschsprachige SAP Anwendergruppe e.V. (*DSAG*)[2] is the German-speaking user group. They plan multiple events throughout the year in Germany. Similarly, Americas' SAP Users' Group, the *ASUG*,[3] operates in the United States and hosts events there.

UI5con

UI5con[4] is a free, community-driven conference focused on the UI5 framework and related technologies. UI5con is typically held in Germany, but it was also held in other locations in the past. The conference is an opportunity for professionals to learn about the latest developments in the frontend field and listen to experience reports from fellow developers. The content covers all sorts of topics, starting at the ideal setup of the development environment, over the build and test pipeline during development, to the final packaging and deployment to production.

Beyond the pure content of the event, this is also an excellent opportunity to connect with other professionals using the framework. Like other grassroot events, anyone can submit proposals for a session – regardless of job title and employer. This guarantees a wide range of topics and helps the entire ecosystem grow (Figure 7-1).

---

[2] https://dsag.de/
[3] www.asug.com/
[4] https://openui5.org/ui5con/

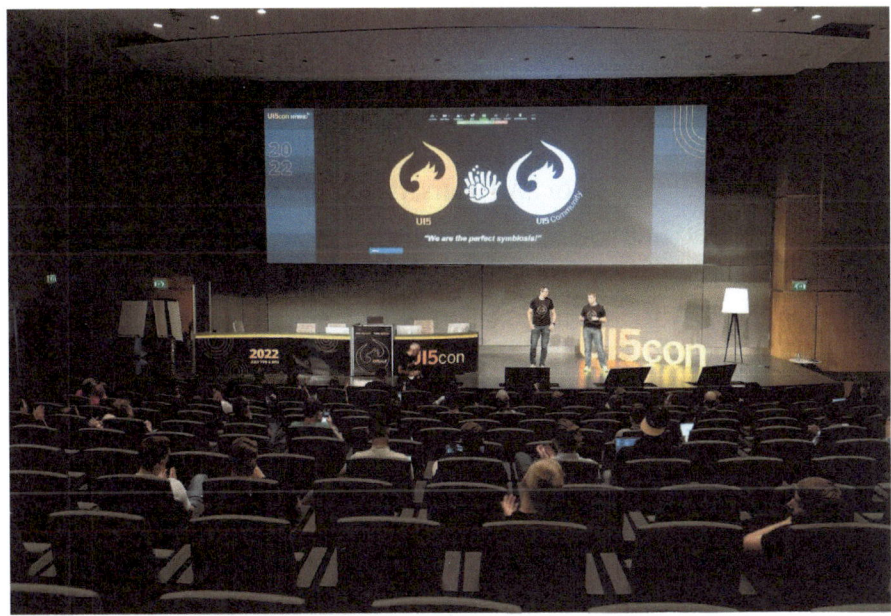

Figure 7-1. *UI5con Hybrid 2022 in St. Leon-Rot*

SAP Stammtisch

An SAP Stammtisch (German for the "regular's table") is a community of SAP users who come together to share knowledge, experiences, and best practices related to SAP software. These informal, regional meetups typically provide a forum for users to connect, discuss SAP-related issues, and learn about new products and technologies.

There are numerous Stammtisch worldwide: Frankfurt, Sydney, Hyderabad, and Paris. There is no fixed cadence or format for when these groups meet and where. Some might have a regular virtual meeting with short lightning talks, while others meet only every once in a while in person for a workshop. A full calendar view that shows when a Stammtisch meets where can be found in the SAP Community[5] or on Twitter.[6] Is there no Stammtisch in your area, but you know there are many SAP developers in the vicinity? Well, then, this might be an excellent opportunity for you to start a new Stammtisch!

These events are open for SAP users, developers, consultants, and architects and don't focus on a single technology and cover anything from frontend development with SAP Fiori to SAP HANA database systems. Depending on the local industry, a Stammtisch might focus on one technology or the other, but the organizers are usually very keen to add new topics to their meetings (Figure 7-2).

Figure 7-2. *SAP Grassroots Stammtisch Frankfurt*

---

[5] https://groups.community.sap.com/t5/sap-stammtisch/eb-p/stammtisch
[6] https://twitter.com/SapStammtisch

SAP Inside Tracks

An SAP Inside Track is a free, community-driven conference that brings together SAP professionals, developers, and enthusiasts to share knowledge and best practices, network, and learn about new technologies and innovations in the SAP ecosystem. Similar to meetups, SAP Inside Tracks are a superb opportunity for SAP professionals to learn from experts in the field, exchange ideas and experiences with their peers, and stay up to date with the latest developments. They are held in various locations worldwide and are open to anyone interested in attending.

The events are usually one-day events on the weekend and are organized and hosted by local communities or a Stammtisch. They often feature presentations, demos, and workshops on various topics related to SAP technologies, such as frontend development with SAP Fiori, the SAP Business Technology Platform, or SAP HANA. Sometimes, SAP Inside Tracks are combined with an SAP CodeJam, a hands-on workshop led by an SAP employee around a contemporary topic that moves the SAP community. SAP CodeJams and SAP Inside Tracks are free and get promoted in the community calendar[7] (Figure 7-3).

---

[7] https://groups.community.sap.com/t5/sap-stammtisch/eb-p/stammtisch

Figure 7-3. *SAP Inside Track Brisbane, AUS*

Joining the Community Virtually

Virtual events got a massive push when the COVID-19 pandemic hit in early 2020, and everyone had to go virtual. We saw, over time, many people wanted to return to in-person events; nevertheless, virtual events are still more common than before the pandemic. This is especially helpful for developers who don't work in metropolitan areas or large companies that can afford to send their employees on business trips.

Online communities are, in general, an essential resource for all developers. They provide multiple learning opportunities, mostly free of charge. First, online communities offer a convenient place for developers to connect and share knowledge and expertise at any time. Second, online communities can provide developers with support and guidance on technical difficulties or challenges they may face. Third, online communities can help developers stay up to date with the latest trends and developments synchronously, in a webinar, or asynchronously while watching a recording.

There are many types of communities and platforms to choose from, which can be overwhelming at first. Many of these communities make use of multiple, overlapping platforms. For this reason, this section is organized by platforms instead of communities. The important part is that you don't feel pressured to join them all at once. It's perfectly fine to focus initially on the communities and platforms you find the most interesting and then expand from there.

SAP Community

The SAP Community,[8] with a capital C, refers to a platform and not a community, even if the name suggests otherwise. It is hosted and moderated by SAP's Community team. Members of the SAP Community can participate in discussions, ask and answer questions, contribute content to the blog, and connect with other SAP professionals and experts. The platform is gamified to reward active users who publish useful content or provide helpful answers to others.

The SAP Community is the go-to resource for multiple events and programs such as SAP Mentors, SAP Champions, first-party events run by SAP, as well as third-party events run by the community like upcoming SAP Inside Tracks or SAP CodeJams.

Furthermore, it offers a variety of resources, including blogs, forums, tutorials, and other learning materials, that are designed to help SAP users and professionals stay up to date with the latest developments in the SAP ecosystem. The SAP Community is organized around different topics and areas of interest, such as SAPUI5, SAP Fiori, SAP S/4HANA, SAP Business Technology Platform, and SAP HANA, among others. The SAP Community addresses all personas in the SAP ecosystem, including UI developers.

---

[8] https://community.sap.com/

It's worth creating an account on this platform and stopping by occasionally to keep an eye on current developments. While the content is moderated, it is worth checking whether a blog post or tutorial is up to date and whether the author is a domain expert or just sharing their personal views on a topic.

YouTube

YouTube[9] is a vast resource for anyone looking to learn new technologies and programming languages, including developers. It's great for developers looking to learn new skills or stay up to date with industry trends. Many YouTube channels offer tutorials on various technical topics, such as programming languages, frameworks, and tools. These tutorials can be helpful for developers looking to learn new skills or to brush up on their existing knowledge. Developers can also find demos and presentations that show how to use specific technologies or tools. These can be useful for understanding how a particular technology works and what it can do. Further, the video platform is a great place to find interviews with technical experts and professionals, as well as panel discussions from various events. These can be an excellent way to learn about different perspectives and approaches to technical problems. Another beneficial category of videos is recordings. Many conferences and events in the SAP tech industry are recorded and available on YouTube. This is especially helpful for developers who want to get a first impression of an event before traveling there.

Because of its enormous collection of uncurated content, it is worth checking whether the content is up to date; if the author is, in fact, an expert; and what other users say about this content in the comment section below the video.

---

[9] www.youtube.com/

- SAP Developers[10]

  SAP-owned channel that posts videos for all developers working in the SAP ecosystem

- SAP TechEd Channel[11]

  SAP-owned channel that posts promotional content for and session recordings of SAP TechEd

- OpenUI5 Channel[12]

  SAP-owned channel that posts content about OpenUI5 and related technologies, including session recordings from UI5con and the monthly webinar "UI5ers live"

- SAP Build Apps[13]

  SAP-owned channel for low-code/no-code content

- SAP Community[14]

  SAP-owned channel covering all sorts of products and technologies for all job roles in the SAP Community

- SAP Inside Track Recordings[15]

  Community-owned website that lists the channels and videos that contain the recordings of many SAP Inside Tracks

---

[10] www.youtube.com/@sapdevs

[11] www.youtube.com/@SAPTechEdChannel

[12] www.youtube.com/@ui5videos

[13] www.youtube.com/@SAPBuild

[14] www.youtube.com/@SAPCommunity

[15] https://opensit.net/

- Code with Brandon[16]

  Community-owned channel with DIY content for
  SAP developers with a focus on UI5

OpenSAP

OpenSAP[17] is an online learning platform by SAP. It offers free massive open online courses (MOOCs) on topics related to enterprise software and technology. The courses are self-paced and designed for professionals who want to gain new skills and knowledge in the SAP ecosystem. OpenSAP courses cover many topics, including SAP Fiori, SAP S/4HANA, and SAP Business Technology Platform. They are developed and delivered by SAP experts and provide learners with videos, readings, and interactive exercises. The platform also offers online forums for each course, in which you can exchange ideas with the staff or fellow students. Upon completing a course, learners can earn a certificate of completion.

OpenSAP is a helpful resource for professionals looking to learn about SAP products and technologies and gain valuable skills that can help them advance in their careers.

Popular courses for frontend developers are

- Evolved Web Apps with SAPUI5[18]

  Course from 2019 covering various topics about UI5

- SAP Fiori Overview: Design, Develop and Deploy[19]

  Course from 2020 covering the SAP Fiori design
  system and how to use it

---

[16] www.youtube.com/@brandcaul/videos
[17] https://open.sap.com/
[18] https://open.sap.com/courses/ui52
[19] https://open.sap.com/courses/fiori3

- Developing and Extending SAP Fiori Elements Apps,
  SAP Fiori Experts[20]

  Course from 2021 covering how to build SAP Fiori
  elements apps and extend an SAP S/4HANA app

Twitter & Co

Twitter[21] is a popular social media platform that allows users to share short messages, called *tweets*, with each other. Tweets can include text, images, videos, or links to other content and can be up to 280 characters in length. The platform also offers a range of features, such as the ability to search for tweets by keyword and create and join groups called "lists."

Twitter is popular among developers because it provides a way to connect, share their work and ideas, and stay up to date on the latest trends and developments. Developers can follow established experts to get access to valuable knowledge and expertise. Experts recognized by SAP, like SAP Mentors and SAP Champions, have a profound understanding of various technologies and often share helpful links and tips with their followers. They frequently share their thoughts and opinions on new technologies, best practices, and other topics relevant to developers working in the same field. Furthermore, experts also share relevant information about upcoming events in their region, such as SAP Inside Tracks, SAP CodeJams, or online content in the SAP Community. Great experts also engage with their followers and answer their questions. This can provide valuable opportunities for beginners to learn from others and gain new perspectives on their work.

---

[20] https://open.sap.com/courses/fiori-ea1

[21] https://twitter.com/

In addition, following project accounts gives developers access to information about specific projects or technologies. For example, a developer interested in a particular framework or open source project can follow the project's account on Twitter to receive updates, learn about new releases, and find out about any opportunities to contribute to the project.

It's a good idea to follow and also interact with a few of the accounts that we're going to introduce in Table 7-1. This will have the benefits that the platform will learn from your actions and start suggesting other accounts you might want to follow and whose content you are likely to find useful.

Table 7-1. *Recommended Twitter accounts to check out and possibly follow*

Twitter Profiles	Descriptions	URLs
SAP Developer Advocates	Lists with individual,	https://twitter.com/i/lists/ 829802253313798145
SAP Champions	recognized experts	https://twitter.com/i/lists/ 11341968831595765777/
SAP Mentors		https://twitter.com/i/lists/ 2351388
SAP Mentors Alumni		https://twitter.com/i/lists/ 1149024118261420032
Best of UI5 Contributors		https://twitter.com/i/lists/ 1551008665313976320

(*continued*)

Table 7-1. (*continued*)

Twitter Profiles	Descriptions	URLs
OpenUI5	SAP Fiori–related	https://twitter.com/OpenUI5
wdi5	projects	https://twitter.com/_wdi5_
Fundamental Library		https://twitter.com/fundamental_lib
Project Luigi		https://twitter.com/luigiprojectio
SAP Business Application Studio		https://twitter.com/sap_bas
SAP Developers	Updates	https://twitter.com/SAPdevs
SAP Community	and content drops from	https://twitter.com/sapcommunity
SAP Mentors	communities that are run by SAP	https://twitter.com/SAPMentors
SAP TechEd	Event updates	https://twitter.com/SAPTechEd
UI5Con		https://twitter.com/ui5con
SAP Stammtisch		https://twitter.com/SapStammtisch
SAP Inside Track		https://twitter.com/SAPInsideTrack
OpenSIT		https://twitter.com/opensitnet
Marius	The authors of	https://twitter.com/iobert_
Volker	this book	https://twitter.com/vobu

At the time of writing, there are multiple controversies around Twitter, and several high-profile individuals publicly announced they want to leave the platform. Hence, we would like to mention a few contenders.

Mastodon[22] is an open source microblogging platform similar to Twitter. It allows users to share short messages, called "toots," with each other and engage in discussions with other users. Mastodon is often used as an alternative to Twitter by users looking for a more privacy-focused or decentralized social media platform.

LinkedIn[23] is a professional networking platform that allows users to connect with other professionals, share their work and achievements, and find job opportunities. It is a platform used by jobseekers, recruiters, and companies to connect and share information about job opportunities, professional skills, and industry trends.

Slack & Co

Slack[24] is a workplace messaging and collaboration platform. Even though it was initially designed for internal communication, Slack is today a popular community platform among developers and other technical teams. It allows users to communicate with each other in real time, share files, and collaborate on projects. Topics can be separated via different channels, and any member of a workspace can join and participate in these individual channels. The platform offers a range of features, including group and private messaging, audio and video calling, and the ability to integrate with other tools and services. One of the key advantages of Slack is its ability to integrate with a wide range of other tools and services. This allows developers to use Slack as a central hub for all their

---

[22] https://mastodon.social/

[23] www.linkedin.com/

[24] https://slack.com/

communication and collaboration without having to switch between different apps and services. Developers can also use Slack to share code snippets, discuss project updates, and receive notifications from other tools like Continuous Integration servers. And therefore, it can be an informal way to get in touch with other community members or even engineers of the frameworks you are using. There are the following Slack workspaces, or channels, that are dedicated to SAP web technologies:

- OpenUI5 Slack[25]

  Workspace with various channels for almost any topic around UI5, SAP Fiori elements, UI5 Web Components, development tools, the UI5 community, and more

- SAP Mentors & Friends (channel)[26]

  Workspace with various channels for grassroot events, almost all SAP products and technologies that are not covered by other workspaces

- Fundamental Library Slack[27]

  Workspace with channels to discuss the Fundamental Library Styles project, including all framework-specific implementations

Discord[28] is another online communication platform that is often used by teams, communities, and organizations to communicate and collaborate.

---

[25] https://ui5-slack-invite.cfapps.eu10.hana.ondemand.com/

[26] https://sapmentors-slack-invite.cfapps.eu10.hana.ondemand.com/

[27] https://join.slack.com/t/ui-fundamentals/shared_invite/zt-6op8woeb-0
~uRqrGZeMm3updfQehbaw

[28] https://discord.com/

It offers features such as voice and video calling, screen sharing, and the ability to create and join multiple channels within a server. Discord is popular among gamers and other communities needing a place to communicate and interact.

Stack Overflow

Stack Overflow is a popular online community for developers to ask and answer questions related to programming, share their knowledge and expertise, and learn from others in the community. Stack Overflow was founded in 2008 and has since then become one of the most popular websites for developers, with millions of users worldwide. The website is organized around the concept of a "stack" of information, where each answer builds upon the previous ones to provide a comprehensive solution to a specific problem. Users can ask and answer questions, upvote or downvote answers, and add comments to help clarify or expand upon an answer. The website also has a reputation system that rewards users who contribute high-quality content and participate in the community. Stack Overflow is a valuable resource for developers because it provides a wealth of knowledge and expertise on various topics – whether you're a beginner looking for guidance on your first project in the SAP UI universe or an experienced developer seeking answers to complex problems. Stack Overflow has a large and active community of users who contribute to the website. This community helps to ensure that the content on the website is accurate, up to date, and relevant to the needs of the developers who use it.

Stack Overflow tags are keywords or labels that can be added to questions to identify the topic or technology being discussed. These tags help organize the content on the website and make it easier for users to find relevant information. When a user posts a question on Stack Overflow, they can add one or more tags to describe the topic of their question. For example, if a user is asking a question about the JavaScript programming language, they might add the "JavaScript" tag to their question. This helps other users who are interested in answering questions about JavaScript to find them more easily. Tags also make it easier for users to search for specific topics on Stack Overflow. The website has a search bar that allows users to enter keywords or tags to find questions and answers related to those topics. For example, a user who wants to see questions about the SAPUI5 programming framework can enter the "sapui5" tag in the search bar to find all questions tagged with that keyword.

As of today, the following Stack Overflow tags exist for SAP-related web technologies:

- sap-fiori[29]
- sapui5[30]
- ui5-webcomponents[31]

Mailing Lists

A mailing list, or announcement list, is an electronic mailing system that allows a group of people to communicate with each other by email. It is typically used to send announcements, updates, or other information to a large group of people who have subscribed to the list. To participate in a mailing list, a person subscribes to it by providing their email address. Once they are subscribed, they will receive emails sent to the list. They can also reply to these emails or send their own messages to the list.

This might be especially interesting for users who want to receive updates directly in their inbox, rather than other platforms that sort their feed algorithmically.

---

[29] https://stackoverflow.com/questions/tagged/sap-fiori
[30] https://stackoverflow.com/questions/tagged/sapui5
[31] https://stackoverflow.com/questions/tagged/ui5-webcomponents

You can subscribe to the UI5.Announce Mailing List[32] to receive messages from the UI5 development team. The list will keep you posted about announcements of new releases, significant changes, and other important news about the UI5 framework and its projects.

GitHub

GitHub is a web-based platform that allows developers to store and manage their code repositories and collaborate with other developers on projects. It uses git, the version control system we also used in this book, to manage code repositories and track changes to the code over time. Developers can, for example, use GitHub to create a repository for a new project, invite other developers to contribute, and track changes to the code as the project progresses. GitHub offers a range of features to support repository-related collaboration and communication among developers. Issues allow developers to track and discuss specific tasks or ideas related to a project. Wikis are used to create and edit collaborative documentation within a repository. And pull requests are the mechanism to submit code changes to a project and collaborate with the project maintainer on the changes before they are merged into the main branch. These and other features make it easier for developers to work together and ensure that their code is of high quality.

In addition to its core features for hosting and collaborating on code, GitHub is also great for community building because it provides a place for developers to connect with each other and share knowledge and expertise. One way that GitHub supports community building is through its support for open source projects. It makes it easy for developers to collaborate on open source projects, which advances the projects and is a great learning

---

[32] https://listserv.sap.com/mailman/listinfo/ui5.announce

and networking opportunity. Further, the platform helps developers to build and strengthen their reputation in the community around those projects. Another way that GitHub supports community building is through its support for social features, such as the ability to follow other users, star repositories, and participate in discussions. These features make it easy for developers to connect, learn from each other, and support each other in their work.

One creative way to use GitHub is for "collaborative documents." An author can create a repository for the document, invite other users to collaborate on it, and even wait for strangers to contribute to it. The version control features of GitHub help track the changes to the document.

We recommend starting to watch a few of these repositories and organizations, which can be seen as collections of repositories. It's fine to be a silent observer in the beginning. And over time, when you become more familiar with technology, why not open an issue when something is broken or engage in a conversation? Or even open your first pull request to one of these interesting projects:

- You can find the code of the open source technologies we covered in this book here:

  - OpenUI5[33]

  - UI5 Web Components[34]

  - Project Luigi[35]

  - Fundamental Library Styles[36]

---

[33] https://github.com/SAP/openui5

[34] https://github.com/SAP/ui5-webcomponents

[35] https://github.com/SAP/luigi

[36] https://github.com/SAP/fundamental-styles

- These are the repositories and organizations that provide tooling support for the abovementioned frameworks:

  - UI5 Tooling[37]

  - easy-ui5[38]

  - UI5 TypeScript Support[39]

  - UI5 Community[40] with tooling that is provided by UI5 developers for UI5 developers, including projects like wdi5[41] and the ui5 ecosystem showcase[42]

  - UI5 Migration[43]

- The SAP Samples[44] organization that includes sample code of demo applications that run out of the box.

- There are also repositories and organizations that contain collaborative documents to share knowledge:

  - SAP Tutorials[45] from the developer center

  - SAP Fiori OData Vocabulary[46]

---

[37] https://github.com/SAP/ui5-tooling

[38] https://github.com/SAP/generator-easy-ui5

[39] https://github.com/SAP/ui5-typescript

[40] https://github.com/ui5-community

[41] https://github.com/ui5-community/wdi5

[42] https://github.com/ui5-community/ui5-ecosystem-showcase

[43] https://github.com/SAP/ui5-migration

[44] https://github.com/SAP-samples

[45] https://github.com/sap-tutorials/Tutorials

[46] https://github.com/SAP/odata-vocabularies

- SAP Tech Bytes Inventory[47]

- UI5 Best Practices[48] by the DSAG

- SAP Documentation[49] of selected products and technologies

- This repository[50] contains all the source code of the examples used throughout this book.

The Journey Just Started

What a ride! We covered quite some ground in this book!

We have discussed why most enterprise software used today looks the way it does and appears a bit behind the consumer-grade apps of large tech companies we interact with daily. But we also started to see that the technologies from these two worlds are not mutually exclusive anymore, and open source technology is making its way into enterprise software. With this, enterprise software companies not only bet on open source but also contribute to it and even publish their open source projects. SAP started doing this as well.

In combination with SAP Fiori, the design system for enterprise user interfaces, SAP now provides multiple options to build the enterprise user interfaces of tomorrow – on any device and platform out there.

---

[47] https://github.com/SAP-samples/sap-tech-bytes

[48] https://github.com/1DSAG/UI5-Best-Practice

[49] https://github.com/SAP-docs

[50] https://github.com/Apress/SAP-UI-Frameworks-for-Enterprise-Developers-by-Marius-Obert-Volker-Buzek

SAPUI5 and its open source counterpart **OpenUI5** are matured frameworks that leverage the MVC pattern. Both try to find the ideal trade-off between reusability and flexibility to build enterprise applications that work seamlessly in the SAP Fiori launchpad.

SAP Fiori elements extend SAPUI5 applications with prebuilt UI elements that implement established UX standards. This increases the development speed enormously and reduces the maintenance effort to a minimum. But it also comes with a cost of flexibility and performance.

UI5 Web Components, on the other hand, provide less abstraction than UI5 and follow, in some way, the opposite approach of SAP Fiori elements. They build on new, standardized technologies to provide reusable components that can be used with any SPA framework or with plain browser features. One of the features of the Web Components technology is that they are reusable but also not very extendable beyond the configuration options the components offer.

This is where **Fundamental Library Styles** comes in. It provides the SAP Fiori look and feel via CSS classes but doesn't come with any behavior. This behavior needs to be implemented by the developers, which gives them the freedom to extend components as needed and easily design new ones. There is also a wrapper for Angular apps to accelerate these development efforts.

We tried to keep this book as timeless as possible, which wasn't easy in this fast-moving field. These chapters will, eventually, become antiquated. Therefore, it's even more critical that you keep an eye on new, trending frontend technologies, especially the ones from the consumer technology sector. They might make the jump into the enterprise software universe in a relatively short time. For this reason, we also briefly identified trends we see on the horizon of frontend development today, followed by an introduction of SAP projects related to these trends.

And finally, we found events and communities you can join to continue your learning journey. They also offer ways to connect with peers around the world – virtually and in person. This list can never be complete, as new communities and platforms arise all the time. But we hope this gives you a starting point to find valuable events and communities, inside and outside the SAP domain, in your area. Now it's up to you to join the communities you like to become a successful developer who builds the user interfaces the SAP universe deserves.

We hope you enjoyed reading this book and consider it a valuable resource!

Index

A

A/B testing, 3
Americas' SAP Users' Group (ASUG), 272
AngularJS, 20, 41
Announcement list, 287
Auditability, 8

B

Backbone.js, 20, 41
Bootstrap, 49, 107, 109, 137, 160, 162, 168, 177, 196, 208
Business Technology Platform (BTP), 95, 240, 275, 277, 280

C

Capital expenditure (CapEx), 13
Cascading Style Sheet (CSS), 208
    Fundamental Library Styles, 210
    SAP Fiori CSS, 221
    and Sass, 210, 229
Cloud computing, 3, 270
CommonJS, 246
Community calendar, 275
Community-driven projects, 263
Community-owned website, 279

Consumer-grade software, 4, 5, 7, 10, 12, 14
Consumer tech companies, 5
Content Delivery Network (CDN), 95, 100, 168
Continuous Deployment (CD), 39, 112–114, 246
Continuous Integration (CI), 103, 112, 114, 285
Continuous Integration and Delivery, 19
Continuous Integration/Continuous Deployment (CI/CD), 26
Cordova applications, 6
COVID-19 pandemic, 270, 271, 276
Cross-site request forgery (CSRF), 25
Cross-site scripting (XSS), 25
CSS mixins, 211, 234
CSS variables, 206, 210, 220
Customization, 2, 8, 12, 192, 228, 239–243

D

Design stencils, 29
Design thinking approach, 3
Developers, 7, 13–23, 26, 38, 40, 41, 43, 84, 86, 88, 125–130, 145

H

Human resources (HR), 7

I, J

K

L

M